TRANSACTIONS

OF THE

AMERICAN PHILOSOPHICAL SOCIETY

HELD AT PHILADELPHIA

FOR PROMOTING USEFUL KNOWLEDGE

VOLUME XXII—NEW SERIES

PART I

Philadelphia:
THE AMERICAN PHILOSOPHICAL SOCIETY
104 SOUTH FIFTH STREET
1911

ARTICLE I.

175 PARABOLIC ORBITS AND OTHER RESULTS DEDUCED FROM OVER 6,200 METEORS.*

(Read April 21, 1911.)

By Chas. P. Olivier.

The first observations on which the results in this paper depend were made on November 14, 1898. This does not include a few records found in some old books, which had been made many years previously, but never apparently used. No year from 1898 to the present has passed without the addition of quite a number of meteor observations until, up to the end of 1910, about 6,000† had been recorded. From lack of experience, both in meteor observing and other lines of astronomical work, the three Leonid, one Perseid, and the Beilid radiants deduced from the 1898 and 1899 observations, can not be considered as accurate.

Even the paths of meteors plotted in 1900 are probably not so good as those since obtained. However, from 1901 on, while naturally each year should improve the methods and accuracy slightly, yet there is reason to feel almost equal confidence in the results.

I am under deep obligation to Director Ormond Stone of the Leander McCormick Observatory, for continued encouragement and advice in this work from its very beginning to the present.

In scarcely less degree am I under obligation to Director Campbell of the Lick Observatory, for encouragement in this work and allowing me time to carry most of the computations to a conclusion while I was a member of the staff there during 1909 and 1910.

It must also be stated that about 1,900 meteors, or nearly one third of the total, were observed by me while at the Lick Observatory. All of the remainder, except about 300, were observed at or near the Leander McCormick Observatory, University of Virginia.

* Presented to the Faculty of the University of Virginia in partial fulfilment of the requirements for the degree of Doctor of Philosophy.

† So far as I have been able to find from records Corder, Denning, Heis, and Zezioli are the only four observers who have each observed over 5000 meteors.

5

I am further indebted to the following observers, who at various times have either assisted in recording or made separate observations at other places under the same general plan. Messrs. T. B. Lyons, G. F. Paddock, K. S. Patton, J. B. Smith and J. P. Smith, of the University of Virginia, Dr. S. Albrecht, Evanden, Miss E. Glancy, Messrs. P. W. Merrill, K. Lows, R. Young, of the Lick Observatory. Occasional or remarkable meteors have been reported by others.

The methods of observing have evolved with increasing experience, but from 1900 on they have not changed greatly. At present a meteor is observed as follows: Maps are prepared of the region of the sky that is to be especially observed on a given night, care being taken to choose that map whose projection is best for the region in question. In a large recording book a number of columns are ruled, headed as follows (1) Time, (2) Number, (3) Class, (4) Color, (5) Magnitude, (6) Length of Path, (7) Duration in tenths of seconds, (8) Duration of Train in tenths of seconds, (9) Remarks, (10) Serial Number, (11) Accuracy. The designations are mostly self explanatory. (2) gives the number of the meteor for the night, (10) the serial number for the year— filled in later, (11) the accuracy, on a scale of 3, with which the meteor was observed. Beside the plotted path of the meteor on the map is placed the number for the night and later the serial number in ink. The serial numbers are so arranged that the first figure itself gives the year during which the meteor was seen. Thus 1—117 shows that the meteor was seen in 1901, 9—1,136 in 1909, etc. The methods used to obtain the most accurate plot of a meteor's path are as follows: The greatest care was taken to obtain the direction and any one point over which the meteor passed. Often, of course, a meteor's beginning and ending points fall exactly at or very near a convenient star, or at such a distance between two near stars that it is easy to estimate the distance proportionally and accurately. In such a case the direction, determined nearly always by holding up a straight rod so that it appeared to lie parallel to the meteor's path in the sky, served mainly as a convenient check.

But in most cases a meteor neither begins nor ends at a point which is easy to determine. Then by glancing backwards and forwards along the rod the eye can always pick up a star in the same great circle. Also there is scarcely ever any difficulty in finding some one point actually in the path itself. As the eye readily estimates the length of path of a meteor with fair accuracy, the parts in front and behind the chosen point can be estimated instantly, and by means of the other reference point entirely outside the path, the meteor's position can be gotten with great accuracy and speed, compared with other methods. By choosing some point behind rather than in front of the path we also in this method eliminate to a great extent the effects of poor projection, which may be troublesome near the edges of almost any map. However

I wish to state that meteors beginning at a greater distance than 30° from the radiant were never given much weight, and whenever there was reason to suspect that a plotted path was distorted by poor projection the meteor was either given extremely little or no weight at all.

For the strong streams such as the Leonids and Perseids, the radiants could frequently be determined by meteors within 10°.

Other meteors further out but well observed were always used, but the resulting point would generally have been practically the same, had they been omitted.

It is obvious that the short paths near the radiant are most useful in its determination, both because of their nearness and also their low apparent velocity, which permits of the most accurate plotting. They nearly always have trains also, a most material assistance.

If for any reason certain meteors could not be at once plotted their paths were described with such detail that afterwards when put upon maps the results were quite comparable with these plotted at the moment. As far as possible, however, each meteor was plotted where observed.* The usual plan was to work up the result partially the next day so that details could be added, when necessary, while the recollection was fresh. The observations made by others, who have assisted or observed elsewhere for me, were made in the same general way, only the results were left to me to work up completely and the responsibility for the latter rests upon me. As confirming the results of previous observers the following points may be noted as of general interest with regard to the work and results. Most of the meteors were observed after midnight and to the east of the meridian. I have no reason to believe that the south-east or north-east quadrants differed appreciably in the number of meteors seen within them. Those seen before midnight differed much, as a rule, in apparent velocity from those seen after, being much slower. Nearly always there was a marked falling off in numbers before the least trace of twilight appeared. Very slow meteors leave trains, nearly without exception. Meteors with curved or sinuous paths are rare (see tables). In very many cases two or even three meteors travel the same apparent paths within a few seconds. In showers like the Perseids, two or more meteors frequently appear at the same instant. In most cases radiants cover large areas only because of poor observations, since the paths of well observed meteors generally intersect nearly in a point or within a small area. The most notable exception to this rule was on August 10 and 11, 1910.

The physical appearance of meteors, such as color, apparent velocity, etc., while all very useful in assigning a given meteor to the proper radiant, can scarcely ever

* On an average a single meteor was plotted and a full record made in about 40 seconds. However, about 60 per hour would be the most possible to observe fully under usual conditions.

be held as conclusive evidence that it does belong to any given radiant.   Many cases could be cited, especially in August and October, in which meteors have every physical characteristic exactly like numbers of the main stream but come from distant radiants.

Also numerous meteors of the principal streams differ very much from the average member of that stream.   This point is to be especially noted in view of the statement often found that a meteor belongs to a given stream because it looks like the average member of it, though its direction was frequently very poorly determined or perhaps not at all.

Radiants have been found by projecting the plotted paths backwards.   In regard to those for which parabolic elements have been computed this rule was followed.   At least three meteor paths on projection must meet within a circle not more than 0°.5 in diameter.

Any other well-observed meteor whose projected path comes within 1° of the center is accepted and given due weight.   Any other meteor whose projected path comes within 2° of the point may be used, but would be given little weight.   The radiant when finally accepted lies at the weighted center of gravity, as it may be called, of the area enclosed by the projections.   The weight given each projection depends on how near the meteor was and how well observed.

An absolute rule has been made that under no conditions have meteors observed on more than one night been used to determine any radiant.

It is my firm conviction that not following this rule has led many previous observers to catalogue hundreds of fictitious radiants whose presence in our catalogues only hampers the future growth of meteoric astronomy.

Only in the case of a radiant known to be stationary could the combination of meteor paths observed on different nights be justifiable, and stationary radiants must be rare phenomena,* since the meteoric apex moves, on account of the orbital motion of the earth, about 1° per day through the sky and each stream has also its own motion in space.

Since such combinations, in effect, presuppose the existence of stationary radiants, they appear to prove that on the assumption of which they largely owe their own apparent existence.   Had I been willing to combine several nights' work there is little doubt that the number of my radiants could have easily been doubled.   Owing to this precaution largely, I presume, my own work gives little indication of stationary radiation.

Most of the meteors observed came during July, August, October and November.

* Refer to *Mon. Not. R. A. S.*, Vol. XXXVIII, p. 115; *Astr. Nach.*, Vol. XCIII, p. 209; *Bulletin Astr.*, Vol. XI, p. 409–10.

Quite a number have been recorded in January, April and May, and a few in December. The other four months have furnished no results, observations never having been made during them.

It should be stated that only a small fraction of my time given to observing could be devoted to meteor observations, which fact explains why the total number is not larger.

The formulæ and methods for computing the parabolic orbits which follow in the tables were taken from "Die Bahnbestimmung der Himmelskorper" by Julius Bauchinger, which has proved invaluable. In the thirty-fifth chapter of this work the equations will be found in full.

Believing that in some cases the radiants were known more closely than to whole degrees the measurements on the maps were made to tenths of degrees. The transformations from right ascension and declination to longitude and latitude and all the actual computations for the elements were made with 4-place logarithms, the angles being taken out to the nearest minute. However, in the tables of results they were again reduced to the nearest tenth degree, that being quite as accurate as the observations could give.

All obtainable works on meteoric astronomy have been freely consulted and, as will appear later, certain conclusions on the question of stationary radiation have been partially reached by discussion of the work of other observers.

Finally it may be stated that it is hoped at some future time to study the data on which these result are based with a view to the solution of other problems, not taken up or only mentioned in this paper.

THE AQUARID METEORS OF MAY.

These meteors were never seen in sufficient numbers before 1910 to deduce a radiant for them. Indeed, cloudy weather or moonlight had never permitted any of the former attempts to be successful. However, in 1910 good radiants were secured on May 4 and May 11. On May 4, at the Lick Observatory, numerous meteors were seen in the southeast by Dr. Curtis and later by myself. During the half hour before dawn, at intervals when my assistance could be dispensed with, I was able to observe 9 meteors, 6 coming from the Aquarid radiant. This was therefore well determined, as the observations were good. During about the same interval Dr. Curtis saw 4 more Aquarids. The hourly rate must certainly have been quite high. The average length of path for the Aquarids was $10°.5$ and average duration $0^s.64$.

On May 5, my undivided attention was given to meteors from $13^h 23^m$ to $14^h 58^m$, during which interval 13 meteors were observed.   Not more than 2 could have been Aquarids.   On May 11, during an intermittent watch of 35 minutes while in the Crossley dome, just before dawn, 8 meteors, of which 5 or possibly 6 were Aquarids, were observed.   A gold radiant was obtained from them.   The average length of path was 7°.8, and the average duration 0ˢ.55.

On May 13, from 14^h7^m to 15^h20^m, 14 meteors were seen.   Not more than 2 could have been Aquarids.

On May 14, information was given me that 14 meteors had been seen by one person from 14^h45^m to 15^h30^m.   Whether any Aquarids were among them was not stated.

On May 15, one Aquarid was seen.

On May 19, during a continuous watch of one hour before dawn, and an intermittent watch earlier, no Aquarids were seen.

The vicinity of the radiant of possible meteors in the tail of Halley's Comet was examined several times for short periods with the 12-inch refractor and low power giving a large field of view.   No meteors were seen with this instrument.

The following table gives the elements for Comet Halley and the parabolic elements deduced from the observations on the three dates.

	G.M.T.	a	δ	ι	Ω	π	$\pi-\Omega$	q
	1910	°	°	° ′	° ′	° ′	° ′	
Comet Halley.......				163 12	57 16	168 58	111 42	0.5869
Aquarids..........	May 4.97	334.0	−3.4	166 15	44 4	155 6	111 2	0.6770
Aquarids*	May 6.93	337.7	−0.6	163 9	45 58	148 17	102 19	0.6067
Aquarids..........	May 11.99	342.0	−0.6	166 41	50 51	155 7	104 16	0.6297

The connection of the meteors with the comet is quite obvious.   The probable connection of these meteors with Comet Halley was pointed out long ago by Professor A. Herschel[†], but the data on which his conclusions were based was not extensive.

However, in 1910 the most interesting point is the enormous size indicated for this meteoric current.   On May 4.97 Halley was about 63 million miles from the earth, which was at the same time about 6 million miles from Halley's orbit.   On May 11.99 these figures had changed to 34 million and 13 million miles respectively.   In other words the space at least partially filled by meteors connected with this comet was presumably a cylinder of 13 million miles radius.   Nothing could illustrate better the extreme complexity of some of the principal meteor currents than this example of their possible size.   It is to be noted also that some of these meteors

* Computed from data given in *Popular Astronomy*, November, 1910, p. 538.   Observed at Vieques, P. R., by George Hurtnell.   I projected the meteors on the printed chart accompanying the article and found η Aquarii at the radiant.   The computation was made on this assumption which must be nearly true.

† *Monthly Notices R.A.S.*, XXXVIII, p. 379.

preceded the comet by 63 million miles and therefore it is probable greater numbers followed it, but of course no data can be gotten on this point. It is of interest to note the eastward movement of the radiant between the dates given, and further that May 6 seems the latest that these meteors have previously been seen.

There is a curious coincidence between some of the elements of the Aquarids and the main Orionid stream. The mean elements are tabulated here for comparison.*

	Mean Date 1900+	L	ι	q	π	Ω	π–Ω
		°	°		°	°	°
Aquarids (3)........	10 May 7.96	316.3	165.4	0.638	152.8	47.5	105.3
Orionids (9)........	1900 to 1908	116.6	161.4	0.536	113.4	25.6	87.8

One can see how closely the inclinations agree, and also that the perihelion distances do not differ very greatly. The longitudes of perihelia differ about 40°.

The July and August Meteors.

Both July and August are months during which meteors are very numerous and, besides, the great Perseid stream offers exceptional opportunities.

My first observations of the Perseids were made in 1899, and every year since has added considerable data. The radiants deduced show unmistakably the regular shift of the radiant from day to day in the direction of increasing right ascension, and, to a smaller degree, also in declination.

But as elliptical orbits have several times been computed for the Perseids it was thought useless to compute either a single new elliptical one or parabolic orbits for the separate dates. However, in the table the residuals from Denning's ephemeris† are given.

In this table are given in order named: (1) G.M.T. of the middle of observations, (2) L, (3) α, (4) δ, (5) numbers of meteors used to find radiant point, (6)‡ Ephemeris—Observed values in α, (7) Ephemeris—Observed values in δ. A stationary meteor is considered to give a separate radiant, and the two cases of this kind are so tabulated. Altogether 37 radiants are given. The next table gives: (1) Date, (2) time of beginning, (3) time of ending, (4) time actually occupied in observing and recording, (5) total number of meteors, (6) rate per hour, (7) factor of rate,§ (8) corrected rate, (9) number of Perseids, (10) rate of Perseids, (11) corrected rate of Perseids. Approxi-

---

* See also *Popular Astronomy*, August-September, 1910, p. 422, for another mention of these meteors.

† See his "General Catalogue of Meteor Radiants," p. 210.

‡ Denning's positions are assumed as being at Greenwich Mean Midnight. However as his are given by dates and not by L, therefore the positions are not strictly comparable.

§ This factor is taken as 1.0 when the night is clear and free from moonlight, when observing conditions are good and when the horizon is unobstructed. If any unfavorable conditions arise, the factor is lowered in the proportion that it is believed the number of meteors seen was diminished.

mately 3,100 meteors were observed during July and August and results from these from the basis for these tables and results.

G.M.T. 1900+	L	α	δ	s $\rightarrow$	Δα E-O	Δδ E-O	Notes.
	°	°	°		°	°	
1899 Aug. 10.75	47.9	39.5	+56.3	32	+5.8	+0.7	
1900 " 10.82	48.8	45.6	55.0	24	−0.3	+2.0	
1901 " 8.79	46.5	37.7	57.4	9	+5.3	−0.8	
1901 " 9.77	47.5	42.0	54.8	28	+2.2	+2.1	
1902 " 10.82	48.2	40.0	57.0	24	+5.4	+0.1	
1902 " 11.8	49.1	46.4	57.0	26	+0.9	+0.4	
1902 " 11.8	49.1	47.0	56.8	1	−0.3	+0.5	
1903 July 21.76	28.5	23.6	50.0	5	−1.4	+1.7	
1903 " 23.84	30.4	26.0	51.2	6	−1.6	+1.1	
1903 " 28.7	35.4	36.0	55.9	4	−6.0	−2.2	[G.F.P.]
1903 " 28.79	35.4	32.4	54.7	3	−2.4	−1.0	
1903 Aug. 11.77	48.9	44.0	56.2	17	+2.6	+.11	
1904 " 10.79	48.8	43.1	56.0	5	+2.3	−1.1	
1904 " 10.83	48.8	41.0	56.2	11	+4.4	−0.9	[J.B.S.]
1904 " 11.77	49.6	45.2	58.0	6	+1.4	−0.7	
1904 " 11.84	49.7	46.7	57.0	25	+0.0	−0.3	[J.B.S.]
1904 " 12.74	50.7	50.8	57.4	5	−3.0	−0.2	
1904 " 14.82	52.7	54.9	63.4	7	−4.3	−5.3	Perseids??
1904 " 16.74	54.5	53.6	61.4	5	−0.6	−2.8	
1905 " 9.57	47.3	41.1	57.2	10	+3.0	+1.6	
1905 " 11.63	49.4	44.7	55.3	24	+1.8	+1.9	
1906 " 10.75	48.2	41.2	57.0	9	+4.1	+0.0	
1906 " 11.75	49.2	43.4	56.0	5	+3.2	+1.3	
1907 " 4.75	42.2	35.0	55.1	4	+3.2	+0.5	
1907 " 11.74	49.0	45.7	56.2	10	+0.9	+1.1	
1908 " 1.74	40.0	28.6	50.4	5	+6.1	+4.4	
1908 " 1.74	40.0	29.2	52.5	1	+5.5	+2.3	No. 8–049
1909 July 23.91	31.2	24.4	50.9	4	+0.0	+1.4	
1909 Aug. 9.88	47.6	39.1	56.8		+5.3	+0.1	
1909 " 10.93	48.6	40.0	57.0	11	+5.5	+0.1	
1909 " 11.87	49.6	42.4	57.5	44	+4.4	+0.2	
1909 " 13.87	51.5	45.6	57.7	17	+3.7	+0.2	
1910 " 1.93	39.7	31.9	58.6	4	+3.0	−3.8	
1910 " 4.95	42.6	37.6	56.6	4	−0.8	−1.0	
1910 " 6.9	44.5	37.6	55.7	12	+3.1	+0.4	
1910 " 10.93	48.4	41.4	56.4	12	+4.1	+0.7	
1910 " 11.86	49.3	44.3	57.9	31	+2.4	−0.6	

While these tables give the principal results some remarks may be added. The richness of the stream varied greatly from year to year, not only in numbers but in the brightness of the meteors, especially near maximum.

So far as numbers go, 1909 August 11 furnished the finest shower, 338 meteors being seen, of which 223 were Perseids. The radiant areas were larger than usual in 1909. (For a fuller description of my observation of this return see the *Lick Observatory Bulletin*, No. 166.)

In 1901 on August 9, meteors were numerous but on former dates scarce. 1902 gave a good display at maximum. In 1903 Perseids were numerous late in July but moonlight spoiled the first half of August. 1904 gave a good display. The maximum of 1905, as observed at Daroca, Spain, was not a conspicuous one though weather conditions were good. During 1906 and 1908 the maximum came in bad weather, while that of 1907 was not a rich one. On an average a Perseid meteor seen before

Date.	Beginning.	Ending.	Total.	$\frac{s}{\text{Total.}}$	Rate.	Factor.	Rate Cor.	$\frac{s}{\text{Perseid}}$	Rate.	Rate Cor	
	h. m.	h. m.	m.								
1899 Aug. 10	10 40	16 18	83	35	25.3	0.9	28.1	32	23.1	25.7	
1900 July 30	12 20	16 0		20		0.4					Intermittent watch.
Aug. 10	12 0	16 50	290	42	8.7	0.4	21.8	40	8.3	20.8	
15	8 0	11 40		49		0.9		8			Intermittent watch.
17	8 40	10 40	120	25	12.5	0.9	13.4	3	1.5	1.7	
18	8 0	10 55		53		1.0					Intermittent watch.
19	9	11	120	20	10.0	1.0	10.0				
1901 July 28	13 58	16 10	132	23	10.5	0.7	14.4	1	0.5	0.7	
Aug. 7	11 23	12 50	87	10	6.9	0.5	13.8	4	2.8	5.6	
8	11 46	15 36	180	32	10.7	1.0	10.7	16	5.3	5.3	
9	10 31	15 58	327	89	16.3	1.0	16.3	55	10.1	10.1	
1902 July 28	13 12	15 34	99	18	10.8	1.0	10.8	2	1.2	1.2	
Aug. 10	9 21	16 9	285	72	14.9	0.8	18.6	44	9.3	11.6	
11	13 40	16 10	150	96	38.4	1.0	38.4	76	30.4	30.4	
1903 July 20	11 52	13 52	120	9	4.5	0.8	5.6	1	0.5	0.6	
21	11 51	14 10	139	19	8.2	1.0	8.2	11	4.7	4.7	
23	14 0	16 0	120	25	12.5	0.7	17.9	6	3.0	4.3	
24	13 0	15 35	155	31	12.0	0.7	17.1	5	3.9	5.6	
27	11 40	15 34	234	80	20.5	0.5	41.0	6	1.5	3.0	
28	11 50	15 55	234	47	12.1	0.7	17.3	3	0.8	1.1	
28			205	33	9.7	0.7	13.9	5	1.5	2.1	[G.F.P.]
Aug. 11	11 10	15 47	174	22	7.6	0.4	19.0	21	7.2	18.0	
11	12 0	14 24	150	18	7.2	0.5	14.4	2	0.8	1.6	[G.F.P.]
1904 Aug. 5	15 10	16 21	71	9	7.6			7	5.9		
10	12 50	14 50	120	21	10.5	0.3	35.0	19	9.5	31.7	
10	12 22	15 22	180	102	34.0						[J.B.S.] Charlevoix, Mich.
11	9 50	16 26	300	95	19.0	0.5	38.0	69	13.8	27.6	
11	13 56	14 40	34	17	30.0			16	28.2		[J.B.S.] Charlevoix, Mich.
12	10 10	14 55	210	26	7.4	0.5	14.8	17	4.9	9.7	
14	12 45	16 5	200	46	13.8	0.8	17.2	15	4.5	5.6	
16	12 20	14 40	80	10	7.5	0.7	10.7	6	4.6	6.6	
1905 Aug. 9	10 10	15 45	250	77	18.5	1.0	18.5	41	9.8	9.8	at Daroca, Spain.
10	13 40	15 0	80	17	12.8	0.5	25.6	16	12.3	24.6	" " "
11	14 0	16 0	120	96	48.0	1.0	48.0	84	42.0	42.0	" " "
1906 Aug. 10	9 40	15 40	194	42	13.0	0.5	26.0	28	8.7	17.4	
11	12 20	13 20	60	16	16.0	0.5	32.0	14	14.0	28.0	
1907 Aug. 1	11 40	15 15	115	20	10.5	0.7	15.0	7	3.7	5.3	
4	12 18	15 21	180	40	13.3	0.7	19.0	15	5.0	7.1	
11	9 32	15 21	181	81	26.9	0.6	44.8	57	18.9	30.2	
1908 Aug. 1	11 29	14 20	171	43	15.1			16	5.6		
1909 July 21	13 12	14 42	90	34	22.7	1.0	22.7	6	4.0	4.0	
23	12 49	14 19	90	20	13.3	0.8	16.6	2	1.3	1.6	
26	12 42	14 42	120	43	28.7	1.0	28.7	2	1.0	1.0	
27	12 56	14 41	105	37	21.1	1.0	21.1	3	1.8	1.8	
Aug. 9	11 52	13 57	120	47	23.5	0.8	29.4	25	12.5	15.6	
10	12 52	15 38	155	122	47.2	0.9	52.4	102	39.5	43.9	22 more by [E.G.]
11	9 21	16 10	395	338	51.3	1.0	51.3	223	35.4	35.4	
12	9 38	9 49	9	8	53.3	1.0	53.3	8	53.3	53.3	
13	11 20	14 30	190	79	24.9	1.0	24.9	42	13.3	13.3	
11	11 17	11 49	32	64	120						[S.A.] counting only.
11	11 15	12 15	60	134	134						[E.] " "
11	12 3	12 26	23	40	104						[P.W.M.] " "
11	13 49	14 13	24	30	75						[P.W.M.] " "
11	14 13	14 41	28	70	150						[P.W.M.] " "
1910 July 28	11 0	11 58		11		0.7					Intermittent watch.
29	10 34	11 27		7		0.8		2			" "
31	12 50	15 50		24		0.8		7			" "
Aug. 1	12 38	15 38		51		1.0		9			" "
4	13 28	16 1		19		0.9		10			" "
6			190	59	18.6	1.0	18.6	29	9.2	9.2	
8	13 56	15 56		18		0.8		8			" "
10	12 53	15 33	160	100	37.5	0.8	46.9	61	22.9	27.9	
11	9 24	15 47	383	263	41.2	0.8	51.5	204	32.0	40.0	
11	12 36	12 56	20	35	105.0	1.0	105.0				[P.W.M.] counting only.

midnight remains visible $0^s.525$ and after midnight $0^s.385$. These figures are deduced from 393 Perseids observed 1903–1909 inclusive for which the durations were tabulated. They usually leave good trains if the meteor is as bright as the third magnitude. Their prevailing color is red or yellow, few blue or green ones being seen. Other radiants in the neighborhood often furnish meteors precisely like the Perseids themselves, and great care has to be used to keep from misidentification, especially in the case of meteors from near β Persei, which come in some numbers about August 10. This trouble is more serious earlier when the Perseids themselves are no more plentiful than some of the other radiants in contemporaneous activity. On any clear night after July 20, one can be fairly certain of seeing enough meteors to well repay observing, and often enough Perseids to obtain a good radiant for them.

THE OCTOBER METEOR STREAMS.

During this month many rich streams, whose radiants are situated in and near Orion, are in activity. This group was observed with great care because several of the best meteor observers have referred to it as the typical case in which a radiant remains in a practically constant position for quite a long period.

The paper dealing most at length with observations of the Orionids, so far as I have been able to find, is that by W. F. Denning in the *Monthly Notices R. A. S.*, Vol. 56, 74–79. He also treats briefly of them in Vol. 50 of the same publication and in his "General Catalogue of Meteor Radiants."

In all these papers it is stated that the radiant is stationary. Later these papers will be referred to at length.

Date.	Began.	Ended.	Total.	Meteors.	Rate.	Factor.	Rate Cor.	Remarks.
	h. m	h. m.	m.					
1900 Oct. 19	11 39	17 13	334	117	21.0	0.9	23.3	
26	8 39	10 44	125 −	16	7.7 +	1.0	7.7 +	.
01	12 27	16 9	220	63	17.2	0.9	19.1	
19	11 24	16 37	313	83	15.9	1.0	15.9	
02 19	12 13	16 50	225	16	3.5	0.2 ±	17.5 ±	
03 18	13 13	16 16	183	54	17.7	0.8	22.1	
19	11 24	17 38	360	144	18.8	1.0	18.8	31 by [J.P.S.]
04 14	13 22	15 5	100	23	13.8	0.8	17.2	
16	12 23	15 29	160	39	14.6	0.9	16.2	
18	11 8	17 16	360	75	12.5	0.6	20.8	
18	12 0	16 30	270	60	13.3	0.7	19.0	[J.B.S.]
18	12 0	16 40	280	55	11.8	0.7	16.9	[J.P.S.]
05 20	14 16	17 16	180	34	11.3	0.6	18.8	
23	14 25	16 15	110	28	15.3	0.9	17.0	
06 12	12 2	12 58	56	13	13.9	1.0	13.9	
25	12 36	17 6	270	76	16.9	1.0	16.9	
26	14 43	16 13	150	33	13.2	1.0	13.2	
07 15	15 18	16 46	88	20	13.7	0.9	15.2	
08 18	11 12	13 45	150	24	9.6	0.8	12.0	
09 12	14 55	17 5	120	27	13.5	1.0	13.5	
13	13 17	16 52	215	47	13.1	0.8	16.4	
15	11 30	16 0	270	88	16.7	0.8	20.9	
19	13 39	15 19	100	20	12.0	0.6 ±	20.0 ±	
22	12 0	15 50	230	69	18.0	0.9	20.0	
10 8	15 7	16 32		7				Intermittent watch.
13	14 5	15 10		8				" "
25	11	13		11				" "

My observations of this most important group of radiants began in 1900 and, during every October since, some data have been collected bearing upon them. The following table gives the number of meteors observed in October for the year 1900 to 1910 inclusive, on nights when regular observations were made. The columns give from left to right :(1) Date, (2) time of beginning, (3) time of ending, (4) number of minutes actually spent in observing, (5) number of meteors, (6) rate, (7) factor depending on sky, etc., (8) corrected rate. The rates are for one observer.

Seven other observers have assisted in this work. Their assistance was especially valuable in 1904, when J. B. Smith and J. P. Smith on October 18 observed 115 meteors at a station 7 miles southwest from the Charlottesville, Va., station.

The other observers only assisted, as a rule, in counting and recording meteors. Of the 1,279 meteors seen in this month I personally observed 1,075 ±. On working over the maps, on which are the paths of such meteors as were well enough observed to be worth plotting, 64 radiants were obtained, which were considered sufficiently accurately determined to have parabolic orbits calculated for them. The elements will be found in the general table of orbits. In the table of poorly determined or uncertain radiants will be found 7 more.

Of the good radiants, the 55 which fall within the region of the sky shown are plotted in Fig. 1. This figure is purposely drawn on a very large scale so that the radiants could be accurately plotted to tenths of a degree. The 11 radiants that belong to the main stream and all of which were observed on either Oct. 18 or Oct. 19 (when L was between $115°.7$ and $117°.4$) fall within a quadrilateral bounded by $\alpha = 90°.0$ and $\alpha = 92°.1$, and $\delta = +13°.6$ and $\delta = 16°.6$. If No. 112 and No. 113 which were observed by J. B. Smith and J. P. Smith are omitted, leaving the 9 observed by myself, the limits reduce to $\alpha = 90°.1$ and $\alpha = 92°.1$, $\delta = +13°.6$ and $\delta = +15°.9$.

In other words the greatest possible deviation from the mean when all are considered is $\Delta\alpha = \pm 1.°05$, $\Delta\delta = \pm 1°.5$. When the 9 observed by myself are considered this falls to $\Delta\alpha \pm 1°.0$, $\Delta\delta = \pm 1°.15$. These greatest possible residuals give evidence of the probable error of any single radiant determined and how nearly the positions can be relied on.

No. 121 should have been combined at one third weight with No. 122 before the orbits were calculated, but, since that was not done, the positions of the two are so combined and plotted on the map as one point. No. 123 and No. 124 were weighted equally and treated in the same manner. In both of these cases they are undoubtedly identical, having been observed on the same night, however, as two maps were used, two positions were obtained, and to be on the safe side orbits were calculated for each position.

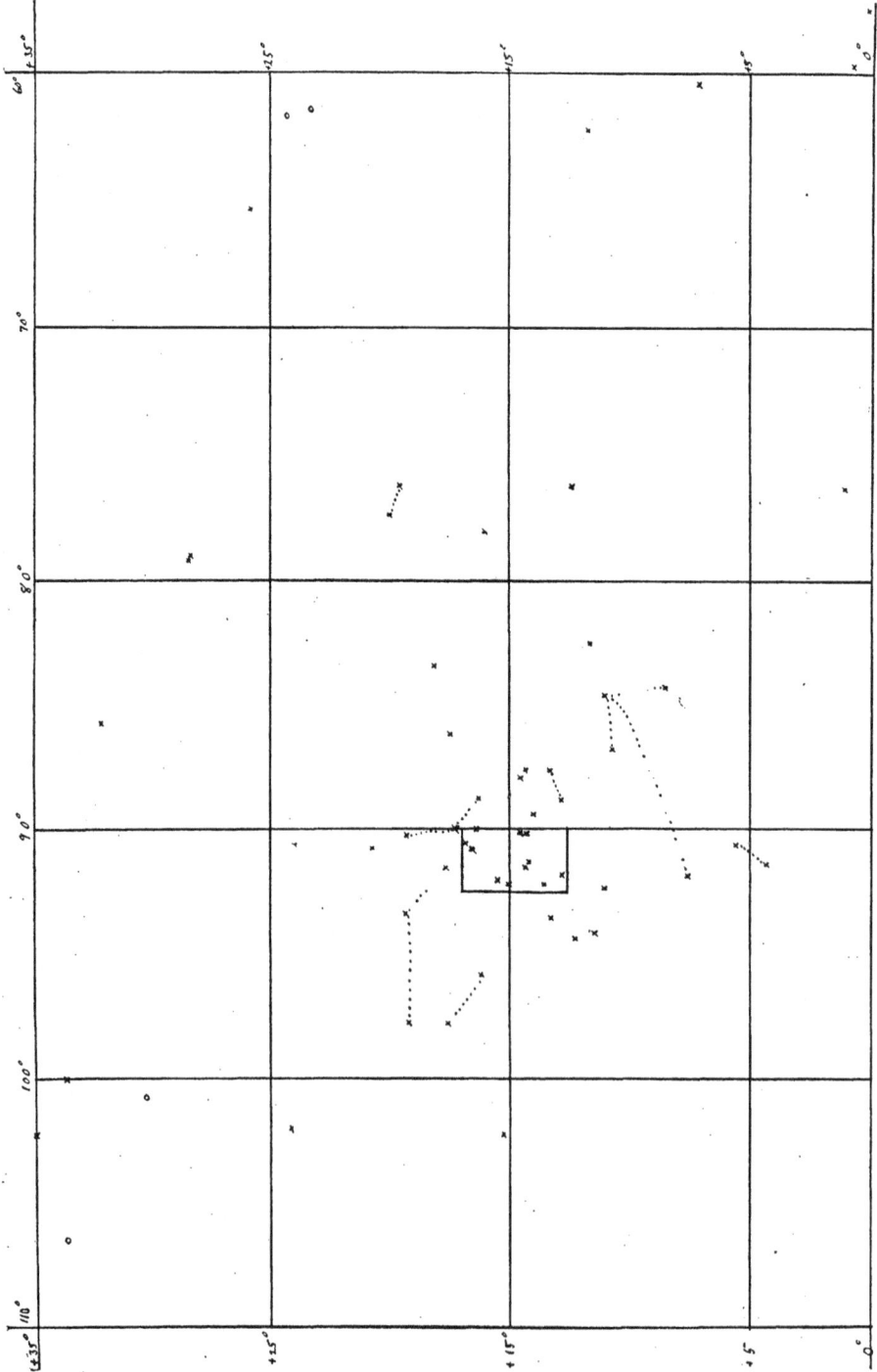

FIG. 1.

The following table groups these 9 positions with regard to the year and L.

No.	Year.	L	a	ϱ	ι	η	π	Ω	q
105	1901	116.2	91.2	+14.2	160.5	130.4	106.0	25.2	0.578
100	1903	115.7	92.1	13.6	159.9	128.0	110.8	24.8	0.618
112	1904	116.4	90.0	16.4	164.6	133.2	111.9	25.5	0.529
113	1904	116.4	90.8	16.6	165.1	132.3	110.2	25.5	0.544
114	1904	116.4	92.0	15.5	161.0	137.4	120.4	25.5	0.456
117	1908	116.5	90.2	14.3	160.2	132.4	110.3	25.4	0.542
Mean.		116.3	91.0	+15.1	161.9	132.3	113.3	25.3	0.544
123 } 124 }	1900	117.4	91.4	+15.4	162.7	132.1	110.2	26.4	0.548
121 } 122 }	1901	117.2	90.7	13.9	159.4	132.7	111.6	26.2	0.541
118	1903	116.6	91.5	14.4	158.7	136.5	118.7	25.7	0.472
Mean.		117.1	91.2	+14.6	160.3	133.8	113.5	26.1	0.520

As the extreme range in L is only 1°.7, and as the difference of the mean α's and δ's fall within the possible errors, it may be permitted to combine the above in ratio 2 to 1, to determine the best parabolic elements for the mean L. Whence we obtain:

$$1900.0 \qquad L = 116.6° \qquad a = 91.1° \qquad \delta = +14.9$$

$$\text{Elements of Orionids} \begin{cases} \iota = 161°.4 \\ \eta = 132°.8 \\ \pi = 113°.4 \\ \Omega = 25°.6 \\ q = 0.536 \end{cases}$$

The following combinations are also suggested:

No.	Year.	L	a	δ	ι	η	q	π	Ω
97	1904	114.4	91.6	+17.7	170.3	127.7	0.628	98.9	23.5
115	1904	116.4	93.4	+19.4	171.8	128.4	0.612	102.2	25.5
126	1905	118.2	97.8	+19.2	172.0	124.4	0.678	96.0	27.3
		116.3			171.4		0.639	99.0	26.1
125	1905	118.2	88.7	+16.2	163.0	138.2	0.443	123.6	27.3
128 } 129 }	1909	120.3	90.0	+17.4	165.3	139.7	0.414	128.8	29.4
130	1905	121.1	90.3	+19.3	169.8	140.0	0.394	132.1	30.3
		119.9			166.0		0.417	128.2	29.0
80	1904	112.4	95.9	+16.1	166.4	118.1	0.776	77.7	21.5
95	1909	113.3	97.9	+17.7	169.8	115.7	0.809	73.4	22.4
		112.8			168.2		0.792	75.6	22.0
87	1909	113.3	76.3	+19.7	170.8	149.8	0.252	132.0	22.4
96	1904	114.4	77.4	+20.0	169.2	156.1	0.163	135.8	23.5
		113.8			170.0		0.208	133.9	23.0
103 } 104 }	1901	116.2	85.7	+10.9	150.4	138.0	0.446	121.2	25.2
98	1903	115.7	84.2	+8.6	144.6	138.7	0.434	122.2	24.8
134	1906	122.9	91.9	+7.7	143.1	137.3	0.457	126.6	32.0
		118.3			146.0		0.446	123.3	27.3
74	1909	110.4	91.3	+4.4	144.1	118.4	0.772	76.3	19.5
78	1909	111.4	90.7	+5.7	145.8	121.3	0.729	83.0	20.4
		110.9			145.0		0.750	79.6	20.0
99	1903	115.7	87.6	+13.3	157.3	135.0	0.498	114.8	24.8
111	1904	116.4	88.9	+12.9	156.6	134.1	0.513	113.8	25.5
		116.0			157.0		0.506	114.3	25.2
73	1909	110.4	42.0	+10.5	17.5	163.3	0.103	166.5	20.4
83	1909	113.3	47.5	+13.1	19.5	165.6	0.062	173.5	22.4
101	1901	116.1	47.8	+11.8	18.8	161.3	0.104	167.8	25.2
		113.3			18.6		0.090	169.3	22.7

However, there still remain within the area bounded by $\alpha = 75°$ to $100°$, $\delta = +5°$ to $+25°$, 13 radiants which apparently do not have any near connection with any group. No less than 8 of these fall within the area bounded by $\alpha = 85°$ to $95°$, $\delta = +10°$ to $+20°$. These 8 were observed on October 13, 15, 18, 19 and 25 of various years.

As stated before the radiants of the main stream which appears on October 18 and 19, all fall within an area bounded by $\alpha = 90°.0$ to $92°.1$, $\delta = +13°.6$ to $+16°.6$. Of the 8 radiants spoken of above only the two seen 1910, October 15, are near enough to the principal radiant for errors of observation to throw them within this area. For the other 6 this possibility hardly exists. Indeed, 3 of these 6 were observed on October 18 and 19, and of the last 3, No. 79 was uncertain, having been gotten from only 3 meteors. To show that the distribution is not entirely without order, even for these isolated cases, it should be noted that no radiant observed after October 19 lies south of $+15°$, except No. 134 and No. 137. No. 137 is too far to the west to enter into the discussion and 134 is evidently connected with No. 103, 104 and 98 in a small system, separate from the main current. I feel quite satisfied that the positions of the radiants given in the tables represent their real places within about $1°$, sometimes less.

Two curious examples of the recurrence of radiants in the same places are given by No. 108 at $\alpha = 79°.2$, $\delta = +28°.6$ on 1904, October 18.81, and No. 132 at $\alpha = 79°.0$, $\delta = +28°.5$ on 1906, October 25.84; also by No. 77 at $\alpha = 87°.9$, $\delta = +14.6$ on 1909, October 13.94 and No. 110 at $\alpha = 87°.5$, $\delta = 14°.4$ on 1904, October 18.81.

The general conclusions drawn from my October meteor observations are as follows:

Within an area bounded by $\alpha = 79°$ to $103°$, $\delta = +4°$ to $+25°$, from October 12 to 26 inclusive, are found a great number of distinct radiants, which in general furnish similar meteors.

That on October 18 or 19 the maxima occur, the principal radiant being always within less than $2°$ of $\alpha = 91°$, $\delta = +15°$.

That minor branches or streams appear which give evidence of an eastward movement in longitude with increase of date.

That these minor streams sometimes appear only during the same October or may reappear in following years.

That many isolated radiants are given which do not seem to have any connection with others either in position or elements.

That since for these radiants an error of as much as $2°$ in the given position seems unreasonable, from a study of the maps and records, they can not belong to one radiant, considered stationary.

Lastly that the suggested explanation is that most of the meteor currents had a common origin, but with the lapse of time have been separated into many minor branches, besides the great central stream.

These minor streams come irregularly, in most cases, and it is by no means necessary to suppose that any given one should appear every year. Indeed it may well be that a small number of meteors give a radiant one year which could never again be observed, because without doubt in the immense extent of such a general system or family of currents many small isolated groups are present which from their small size would never again cut the earth's orbit in future returns to the sun.

In Vol. 56 of the *Monthly Notices R. A. S.*, 74–79, there is an article by W. F. Denning on the Orionids, in which he gives his grounds for concluding that their radiant remains stationary. However, in reviewing it, it is found that in the first table of 19 radiants, 6 are useless for this discussion.* In the second table of 30, 17 are also of no value, the reason being that observations of different nights and years were combined. The 11 available ones in the first table, all observed by Denning himself, fall within an area 3° by 3½°. The 13 in the second fall within 6° by 7°. Therefore what he then called a stationary radiant covered at least 6° by 7°, or were we to discuss all the 49 radiants given by him 8° by 9°. Later in his "General Catalogue" are found 57 radiants assigned to this shower, scattered over an area of 8° by 8½°. When the 25 radiants not observed on a single night are thrown out, the rest lie within 8° by 6°. Nos. 51 and 52 require mention. Deduced from the same observations of Zezioli by Schiaparelli and later Denning, the results differ 2° in R.A. and 3° in decl. Further most or all of the 19 radiants given as Group LXXIX appear to have as much right to be included under the Orionids as many given as such. Group LXXV of 10 radiants, 8 gotten by combining two or more nights' observations, also shows the same coordinates as the Orionids. Were these two groups combined, the Orionids could be made to appear throughout the whole year. Indeed, radiants are very numerous in this region of the sky and stationary radiants can be made to appear by loose combinations of observations and uncritical selection of material. A close study of Denning's own "General Catalogue" from 244 to 250 inclusive will show that many other equally logical conclusions could be reached by merely regrouping. Therefore it is very unsafe to conclude from this data that the Orionids really have a stationary radiant, as stated by him.

THE LEONID METEORS.

Observations on the Leonids were first made on November 14, 1898. Afterwards they were continued in 1899, 1900, 1901, 1903, 1904, 1907 and 1909.

* See Denning's own words p. 78, lines 7 to 10.

Cloudy weather or moonlight prevented observations in 1902, 1905 and 1908. However, in all the other years mentioned meteors were observed, the richness of the showers varying very greatly.

Three tables are given for the Leonids, two quite similar to those already explained for the Perseids. The third is one giving the estimated durations for the Leonids in tenths of seconds. These estimates of course cannot be very accurate, but it is believed the means represent the truth fairly well. Practically all these meteors were observed after midnight. The mean duration for the 257 given in the table is $0^s.39$.

Referring to the table of radiants it will be seen that no certain movement from date to date is indicated. But since L changes only a little over $3°$ for the extreme dates, this is hardly to be wondered at.

Altogether about 1,030 + meteors have been observed between November 12–16 of the years enumerated above.

1901, November 14, furnished the finest shower; 1898, November 14, the second best; while in 1903 and 1904 considerable numbers appeared, with many very bright meteors, particularly in 1903.

Many Leonids give exceptionally long apparent paths and leave splendid trains which remain visible from 1^s to 5^s, often longer. They also furnish bright meteors of several different colors, which would seem to indicate the preponderance of different elements in individual meteors.

Date.	Beginning.		Ending.		Total.	$\overset{s}{\underset{\text{Total.}}{\longrightarrow}}$	Rate.	Factor.	Rate cor.	$\overset{s}{\underset{\text{Leonid.}}{\longrightarrow}}$	Rate.	Rate cor.	Remarks.
	h.	m.	h.	m.	m.								
1898 Nov. 14	13	15	17	40	240±	120	30.0	0.4	75.0	100	25.0	62.5	
1899 " 14	14	11	14	53	42	7	10.0	0.2	50.0	7	10.0	50.0	
1899 " 14	12	30	18	30	360	20	3.3	0.4	8.2	14	2.3	5.7	
1900 " 12	14	13	16	50	157	11	4.2	0.6	7.0	2	0.8	1.3	
1900 " 13	16	40	17	50	70	14	12.0	0.6	20.0	9	7.7	12.8	
1900 " 14	12	12	17	12	150	25	5.0	0.4	12.5	15	6.0	15.0	
1900 " 15	12	20	17	30	235	30	7.7	0.7	11.0	13	3.3	4.7	
1901 " 13	12	30	17	45	300	57	11.4	1.0	11.4	15	3.0	3.0	
1901 " 14	16	50	18	18	88	82	55.9	0.5	111.8	75	51.1	102.2	
1901 " 15	13	10	17	55	265	74	16.8	1.0	16.8	41	9.3	9.3	
1903 " 12	14	35	16	18	103	14	8.2	0.6	13.7	1	0.6	1.0	
1903 " 14	12	28	17	28	275	80	17.5	0.5	35.0	55	12.0	24.0	2 observers.
1903 " 15	12	39	16	58	233	92	23.7	0.4	59.2	78	20.0	50.0	
1903 " 18	10	50	13	20	140	17	7.3	0.6	12.2	1	0.4	0.7	
1904 " 14	12	33	17	23	275	93	20.3	0.9	22.6	65	14.2	15.8	
1904 " 16	14	55	17	16	141	28	11.9	0.6	19.8	10	4.3	7.2	
1906 " 16	11	50	17	21	306	105	17.8	0.8	22.2	52	10.2	12.8	
1907 " 13	14	18	16	18	120	18	9.0	1.0	9.0	1	0.5	0.5	
1907 " 14	14	38	15	58	140	48	20.6	1.0	20.6	3	1.1	1.1	
1909 " 14	12	32	15	52	180	44	14.7	0.9	16.3	14	4.7	5.2	
1909 " 15	14	12	17	36	160	54	20.2	0.8	25.2	10	3.7	4.6	

G.M.T.		L	a	δ	$\xrightarrow{s}$	
1898 Nov.	14.86	142.7	153.0°	+20.8°	29	
1899 "	14.82	142.4	151.4	+19.3	3	
1899 "	15.86	143.5	151.2	+20.6	8	
1900 "	14.82	141.3	151.	+21.		
1900 "	15.84	144.2	151.	+21.		
1901 "	13.86	142.0	150.8	+22.5	13	
1901 "	14.95	143.1	151.6	+21.8	49	
1901 "	15.87	144.0	150.6	+22.4	37	
1903 "	14.84	142.4	151.2	+21.7	20	
1903 "	15.83	143.5	150.8	+22.0	51	
1903 "	15.83	143.5	150.8	+22.1	1	No. 3–562
1904 "	14.88	143.3	150.6	+21.8	34	
1904 "	16.85	145.3	151.6	+22.1	6	
1906 "	16.82	144.8	151.3	+22.4	36	
1909 "	14.84	143.1	149.3	+21.6	7	
1909 "	16.00	144.2	150.6	+23.1	10	

DURATIONS OF LEONIDS.

Date.	1.2	1.0	0.6	0.5	0.4	0.3	0.2	0.1
1903 Nov. 12						1		
" 14				2	9	20	2	
" 15				7	46	19		
" 18					1			
1904 " 14			6	12	30	9	4	1
" 16			1	2	7			
1906 " 16	1	2	2	4	26	13	2	1
1907 " 13							1	
" 14					1	1		1
1909 " 14			1	2	5	2		
" 15				1	6	6	1	
Totals	1	2	10	30	131	70	10	3

The following November radiants are considered to be connected and are given with their elements. The means are given for each group.

No.	Year.	L	a	δ	ι	η	q	π	Ω
150	1904	143.3°	138.0°	+49.2°	122.6°	113.3°	0.834	99.4°	232.5°
158	1900	144.2	138.8	+47.8	124.6	113.7	0.829	100.9	233.5
160	1906	144.8	137.	+49.	122.2	116.3	0.797	106.6	234.1
					123.1		0.820	102.3	233.4
176	1907	142.6	139.5	+35.0	147.6	107.4	0.900	86.7	231.9
147	1909	143.1	139.9	+34.8	147.6	107.6	0.899	87.5	232.3
					147.6		0.900	87.1	232.1
146	1909	143.1	128.6	− 2.0	140.0	120.6	0.733	113.6	52.3
163	1906	144.8	130.7	− 1.2	143.0	120.0	0.741	114.2	54.1
					141.5		0.737	113.9	53.2
152	1903	143.5	72.5	+40.5	43.2	153.0	0.203	178.9	232.8
155	1901	144.0	70.2	+41.5	40.2	150.3	0.242	173.9	233.2
					41.7		0.222	176.4	233.0
142	1901	142.0	159.2	+52.6	112.6	94.3	0.984	59.8	231.3
148	1909	143.1	155.0	+58.0	107.0	94.4	0.993	61.2	232.3
157	1901	144.0	155.4	+50.9	116.3	98.8	0.966	70.8	233.3
					112.0		0.981	63.9	232.3

EXISTENCE OF STATIONARY RADIANTS.

As a typical case of a stationary radiant the group No. XLIII in Denning's "General Catalogue of Radiant Points of Meteoric Showers" was chosen for study. So far as the theoretical impossibility of all the radiants put within this group being really connected is concerned, it is most clearly proved by Th. Bredichin in his memoir "Sur L'Origine Des Étoiles Filantes," pp. 39–44. However, further observational data will be quite useful.

Therefore 68 maps have been examined which contain meteors recorded from 1900 to 1909 inclusive, and on which meteors coming from the region $\alpha - \beta$ Persei would be plotted. These maps were used in January, April, July, August, October and November, all being months during which this shower is supposed to be visible.

On 24 of the maps not a single meteor can be found whose projected path would come within 5° of the point $\alpha = 47°.3$, $\delta = +45°.0$. On the 18 others one meteor might fall within these wide limits but is considered to belong to some other radiant for good reasons. Of the remaining 26 maps, 12 have one meteor each which would satisfy conditions. On the 14 several are found, but in most cases these were used about August 11, and these meteors clearly belonged to the main Perseid stream.

However, three radiants are actually found in this region. No. 33, $\alpha = 42°.4$, $\delta = +49°.2$, No. 38, $\alpha = 43°.8$, $\delta = +39°.6$ and No. 54, $\alpha = 46°.0$, $\delta = +45°.3$. For completeness No. 65, $\alpha = 44°.1$, $\delta = +52°.0$ might be added. The dates on which these were observed are as follows: 1901 August 8 and 9, 1903 August 11, 1904 August 11. For group No. XLIII the limits given are $\alpha = 42°$ to $51°$, $\delta = +39°$ to $+49°$. Therefore three of the above fall within them. In no other month have I been able to confirm the existence of a radiant within these limits.

A little analysis of the data in the "General Catalogue" will be helpful in understanding how such results were gotten. No less than 59 positions are there given. Of these only 15 were obtained from observations of a single night, therefore the remaining 44 are nearly worthless for the discussion of stationary radiation. For example (1) depends on 8 meteors seen within 24 days, (13) on 3 meteors within 9 day, (22) on 11 meteors in 8 days, (50) on 34 meteors during all October and November, etc. Equally bad or worse examples could be quoted, these being taken at random. It should be plain to any observer of meteors or to any person familiar with their theory that such combinations are unsafe and generally misleading. Fortunately out of the 15 positions properly determined, 4 were observed by Zezioli and the resulting orbits calculated by Schiaparelli. His results follow:

D Z+S	Date.	L	a	δ	ι	Ω	π	q
(2) = (4)	Jan. 11	201°	47°	+40°	9°D	291°	131°	0.970
(19) = (137)	Aug. 7	45	42	+48°	53 R	135	341	0.949
(26) = (142)	Aug. 11	49	47	+43	43 R	139	338	0.973
(4) = (150)	Sept. 18	86	51	+39	41 R	176	273	0.561

For purposes of comparison my own four orbits follow:

Ol.	Year.								
(33)	1901	Aug. 8	46°.5	42°.4	+49°.2	126°.5	136°.0	339°.0	0.973
(38)	1901	" 9	47 .5	43 .8	+39°.6	142 .6	136 .9	298 .9	0.989
(54)	1903	" 11	48 .9	46 .0	+45 .3	134 .0	138 .4	295 .4	0.973
(65)	1904	" 11	49 .7	44 .1	+52 .0	122 .9	139 .2	339 .1	0.983

Referring to the four radiants of Zezioli above, we are forced to believe that one of the best of meteor observers plotted 4 radiants over limits of 9° × 9°, when they should have been near the mean position, if we try to obtain a stationary radiant here. But if doubt remains a mere glance at either of the above tables of elements must banish it completely.

That the radiants (137) and (142) of Zezioli are perhaps the same with its position shifted in the 4-day interval is indeed probable. Also my orbits (33) and (65), and (38) and (54) are probably the same, their positions being slightly shifted between the dates. But that all 8 orbits can refer to the same stream is an obvious and mathematical impossibility.

As for the rest of the 15 radiants we had under discussion, (6) seems determined by Denning from a fire-ball path, (7), (34), (38), are found by the same observer from 4 to 6 meteors each. For (14) and (43) no numbers are given. (6), (18), (20), (21), (23) are from duplicate observations of one meteor, and so are not very valuable for the question under consideration.

Therefore of the whole 15 reliable ones, which include (6), (18), (20), (21), (23) also, 7 fall in August. It then follows that we find 8 radiants for all the other months scattered over the area 9° to 10° and it is on these mainly that the claim for an observed stationary radiant here must rest. It is of further interest to remark that these radiants were observed all the way from 1867 to 1897 inclusive.

TABLE OF RADIANTS.

No.	G.M.T.1900+	L	a	δ	l'	b'	l	b	ι	η	q	π	Ω	s→	
		°	°	°	°	°	°	°	°	°	°	°	°		
1	05 Jan.	1.80	191.3	199.4	+47.6	172.2	+50.0	112.0	+74.4	92.8	105.3	0.915	131.9	281.3	3	
2	09	18.83	208.9	174.4	+ 8.4	171.6	+ 5.4	145.6	+ 8.0	162.8	152.0	0.217	142.6	298.6	5	
3	09	18.83	208.9	211.1	+21.7	200.2	+32.0	189.8	+53.6	124.9	101.1	0.948	140.8	298.6	3	D., CLXI, 3, Z15?
4	06 Apr.	18.78	297.4	274.0	+33.2	276.0	+56.5	2.0	+75.1	96.7	76.6	0.951	181.5	28.3	3	Lyrids.
5	04	18.75	297.9	277.0	+36.7	281.2	+59.9	342.5	+76.7	99.7	99.2	0.989	227.1	27.8	4	Lyrids.
6	03 July	21.76	28.5	271.8	+41.2	273.2	+64.6	224.4	+32.5	33.6	103.6	0.960	325.4	118.3	4	
7	09	21.87	29.2	4.6	+31.4	17.6	+26.8	6.3	+44.6	133.2	105.9	0.940	330.7	118.9	4	Compare 10?
8	09	21.87	29.2	348.9	+ 0.9	350.9	- 5.2	323.7	- 7.5	162.6	154.1	0.193	67.2	298.9	4	
9	03	23.84	30.4	0.2	+24.9	10.6	+22.6	353.4	+36.5	137.3	119.7	0.782	359.5	120.1		
10	03	24.81	31.4	2.4	+29.4	14.7	+25.8	359.0	+42.2	133.1	113.2	0.858	347.4	121.1	5	Compare 7?
11	03	24.81	31.4	15.8	+34.8	28.7	+25.8	26.2	+42.9	137.0	93.6	1.012	301.2	121.1	4	Z 110.
12	03	27.79	34.4	2.0	+41.4	21.0	+36.6	2.2	+59.6	116.5	105.5	0.943	335.0	124.0	4	
13	03	27.79	34.4	5.1	+51.3	30.8	+43.9	20.2	+72.6	106.9	94.1	1.010	312.2	124.0	3	D., VII, 10, Z101?
14	03	27.79	34.4	299.9	+49.0	325.9	+66.7	241.8	+49.0	52.4	107.8	0.921	339.6	124.0	7	
15	03	27.79	34.4	307.9	+56.7	349.1	+69.9	238.8	+57.9	60.3	102.9	0.965	329.8	124.0	4	Z 114.
16	03	27.79	34.4	337.7	-16.6	333.2	- 6.7	294.9	- 7.6	40.1	168.2	0.042	100.4	304.0	5	δ Aquarids.
17	03 July	27.79	34.4	339.9	+60.9	23.0	+60.2	250.7	+79.2	81.3	96.4	1.003	316.8	124.0	11	D., CCLXIV, 3*
18	03	27.79	34.4	350.4	+48.2	16.5	+46.7	332.8	+71.8	98.9	105.9	0.939	335.9	124.0	4	Compare 24.
19	09	27.91	35.0	15.4	+15.4	20.1	+ 8.2	9.4	+13.5	165.1	114.5	0.841	353.6	124.6	3	
20	03	28.79	35.4	22.2	+59.5	47.8	+45.6	79.0	+72.9	102.5	78.2	0.973	281.4	125.0	3	D, XV, 19* [G.F.P.]
21	03	28.79	35.4	297.0	+ 7.3	300.6	+27.9	253.3	+18.1	22.6	126.1	0.663	17.2	125.0	5	[G.F.P.]
22	03	28.79	35.4	334.3	-13.7	334.3	- 4.0	296.3	- 4.6	27.9	170.2	0.030	105.4	305.0	12	δ Aquarids.
23	03	28.79	35.4	352.9	+21.4	357.5	+22.4	327.1	+31.9	121.1	141.8	0.388	47.6	125.0	6	[G.F.P.
24	03	28.79	35.4	357.6	+47.7	21.8	+43.6	353.3	+69.8	105.4	103.3	0.962	331.6	125.0	4	Compare 18.
25	02	28.82	35.5	337.6	-15.4	333.4	- 5.6	295.1	- 6.2	32.0	168.2	0.042	101.5	305.1	9	δ Aquarids.
26	01	28.85	35.9	10.3	+45.6	30.1	+37.3	21.4	+62.1	117.2	95.6	1.002	318.6	125.5	5	
27	07 Aug.	1.73	39.5	17.8	+36.8	31.3	+26.8	23.1	+45.0	134.0	101.0	0.978	340.8	128.8	3	
28	07	1.73	39.5	26.9	+53.2	46.7	+38.8	58.4	+64.4	114.3	81.6	0.993	292.0	128.8	4	
29	07	4.75	42.1	18.8	+45.8	36.7	+34.6	29.4	+57.9	121.5	96.5	1.001	324.6	131.7	3	
30	04	5.88	43.9	22.4	+56.5	45.8	+43.0	51.5	+71.4	71.6	87.4	1.012	308.3	133.4	5	
31	01	8.79	46.5	31.2	+31.8	40.2	+18.0	35.0	+30.3	149.3	99.5	0.986	334.8	136.0	2	Compare 37.
32	01	8.79	46.5	32.2	+20.8	37.1	+ 7.3	30.4	+12.3	167.3	105.2	0.944	346.4	136.0	4	
33	01 Aug.	8.79	46.5	42.4	+49.2	55.6	+31.2	27.0	+52.0	126.5	101.5	0.973	339.0	136.0	3	A. P. 14; 21, Z 137.
34	01	9.77	47.5	10.4	+49.0	32 4	+40.2	7.1	+64.6	110.1	106.0	0.937	348.8	136.9	6	D., VII, 20*, 21*
35	01	9.77	47.5	14.9	+27.9	24.8	+19.8	6.3	+31.6	141.0	56.3	0.770	249.6	136.9	3	D, XIII, 5, 4*
36	01	9.77	47.5	40.6	+52.6	55.8	+34.9	66.8	+57.9	120.5	79.6	0.980	296.0	136.9	20	A. P. 14; 26 and 27?
37	01	9.77	47.5	31.8	+32.7	41.0	+18.6	35.7	+31.4	148.0	100.6	0.937	337.9	136.9	3	Compare 31.
38	01	9.77	47.5	43.8	+39.6	53.2	+21.9	58.2	+36.9	142.6	81.0	0.989	298.9	136.9	3	D., XXXV, 3.
39	01	9.77	47.5	349.5	+27.3	2.2	+29.1	323.0	+37.9	97.8	38.4	0.390	213.6	136.9	4	
40	05	9.57	47.3	29.0	+38.5	41.0	+24.8	35.2	+41.8	137.6	81.4	0.991	299.5	136.8	4	D, XXV, 12*, 13*
41	09	9.88	47.6	35.5	+34.9	44.9	+19.6	42.8	+33.2	148.8	93.8	1.009	324.6	137.1	4	D, XXX, 11.
42	09	9.88	47.6	41.4	+26.3	47.0	+ 9.8	46.5	+16.7	163.3	90.8	1.013	318.6	137.1		
43	09	9.88	47.6	42.4	+60.1	60.6	+41.4	84.3	+66.9	108.8	76.3	0.957	289.7	137.1	4	Z 139.
44	02	10.82	48.2	1.1	+60.1	35.4	+52.3	331.1	+80.0	92.4	99.3	0.985	336.2	137.7	4	
45	02	10.82	48.2	327.9	+29.0	342.5	+39.0	288.5	+37.7	57.7	133.7	0.529	45.1	137.7	3	
46	09	10.93	48.6	5.2	+18.1	12.1	+14.5	345.4	+21.2	139.8	145.9	0.319	69.8	138.1	3	
47	09	10.93	48.6	35.0	+56.5	54.0	+39.7	63.3	+66.0	113.2	83.9	1.002	305.8	138.1	10	A. P. 10; 8.
48	04	10.83	48.8	10.2	+55.2	36.8	+45.5	6.8	+73.1	102.9	101.1	0.976	340.3	138.2	3	D, VII, 23.
49	04 Aug.	10.83	48.8	28.2	+31.2	37.4	+18.4	28.2	+30.6	147.8	107.1	0.926	352.4	138.2	5	Compare 31? 37?
51	03	11.77	48.8	5.7	+58.5	37.6	+49.0	21.2	+70.3	107.7	98.8	0.989	335.9	138.2	2	Compare 173, 57.
52	03	11.77	48.9	23.0	+52.4	43.6	+39.2	34.5	+65.2	114.2	95.8	1.002	329.9	138.4	3	D, XV, 34. [G.F.P.]
53	03	11.77	48.9	32.7	+56.2	52.3	+40.0	58.4	+66.8	112.9	94.0	1.009	326.2	138.4	5	A. P. 10; 15. P. 14, 33. Compare 175. [G.F.P.]
54	03	11.77	48.9	46.0	+45.3	56.8	+26.8	64.6	+44.9	134.0	78.5	0.973	295.4	138.4	3	D, XLIII, 23, 26, Z 142. [G.F.P.]
55	02	11.84	49.1	31.6	+35.1	41.8	+20.9	35.5	+35.2	144.1	100.6	0.979	359.8	138.5	4	A. P. 10; 6.

TABLE OF RADIANTS.—*Continued.*

No.	G.M.T.1900+	L	α	δ	l'	b'	l	b	ι	η	q	π	Ω	$\frac{s}{\to}$	
56	09	11.87	49.6	0.2	+26.6	11.4	+24.2	339.8	+34.3	117.5	140.6	0.409	60.1	139.0	4	
57	09	11.87	49.6	7.5	+55.4	35.1	+46.5	357.5	+73.3	100.6	103.0	0.962	345.1	139.0	4	D, VII, 23. Compare 173, 51.
58	09	11.87	49.6	18.4	+25.9	26.9	+16.8	9.3	+26.8	146.7	124.8	0.684	28.5	139.0	3	
59	09	11.87	49.6	43.6	+28.4	49.4	+11.2	49.3	+19.0	171.0	180.0	1.013	139.0	139.0	3	
60	09	11.87	49.6	58.3	+74.3	76.6	+52.3	90.4	+61.8	111.9	71.8	0.914	282.7	139.0	3	D. XLV, 12.
61	09	11.87	49.6	335.6	+68.4	34.6	+66.0	246.9	+68.9	69.8	96.4	1.001	331.7	139.0	4	
62	09	11.87	49.6	339.3	+52.4	11.6	+54.4	293.3	+63.8	78.0	113.4	0.853	5.8	139.0	9	
63	09	11.87	49.6	353.4	+76.4	57.2	+64.5	63.6	+75.4	104.2	86.4	1.011	311.7	139.0	3	D, CCLXXVII, 7*, 8* [J.B.S.]
64	04	11.84	49.7	42.1	+30.4	48.8	+13.5	48.2	+23.0	23.0	89.1	1.013	317.3	139.2	3	[J.B.S.]
65	04 Aug.	11.84	49.7	44.1	+52.0	58.0	+33.5	31.2	+55.8	122.9	100.0	0.983	339.1	139.2	5	Z. 139, [J.B.S.]
66	09	13.87	51.5	15.5	+43.5	32.9	+33.7	10.6	+53.9	119.1	112.4	0.865	5.4	140.9	5	
67	09	13.87	51.5	21.8	+45.7	38.9	+33.6	23.2	+55.2	121.6	105.4	0.942	351.7	140.9	5	
68	04	14.82	52.6	23.2	+21.2	29.3	+10.7	12.4	+17.1	158.2	127.6	0.636	37.2	142.0	3	
69	04	14.82	52.6	35.0	+51.8	51.6	+35.4	50.0	+59.4	120.6	91.0	1.012	324.0	142.0	3	A. P. 10; 13.
70	04	14.82	52.6	46.2	+64.4	65.1	+44.6	11.0	+71.8	103.9	101.8	0.970	345.6	142.0		
71	04	14.82	52.6	54.9	+63.4	69.6	+42.4	105.3	+69.5	102.6	73.7	0.932	289.4	142.0		
72	04	14.82	52.6	73.0	+43.2	76.8	+20.5	97.5	+32.3	137.6	52.3	0.635	245.9	142.0	3	
73	09 Oct.	13.01	110.4	42.0	+10.5	42.7	− 5.4	1.6	− 5.5	17.5	163.3	0.103	166.5	20.4	5	Compare 101, 83.
74	09	13.01	110.4	91.3	+ 4.4	90.3	−19.0	75.8	−31.0	144.1	118.4	0.772	76.3	19.5	3	Compare 78.
75	09	13.94	111.4	62.2	+11.8	62.5	− 9.2	38.7	−11.5	125.2	165.8	0.059	172.2	20.4	5	
76	09	13.94	111.4	87.4	− 3.7	86.7	−27.1	63.1	−42.5	126.5	122.9	0.704	86.2	20.4	4	
77	09	13.94	111.4	87.9	+14.6	88.0	− 8.6	71.1	−14.1	162.0	128.0	0.620	96.4	20.4	5	
78	09	13.94	111.4	90.7	+ 5.7	90.7	−17.8	74.2	−28.8	145.8	121.3	0.729	83.0	20.4	4	Compare 74.
79	09	13.94	111.4	94.4	+12.2	94.4	−11.1	81.8	−18.4	159.2	117.1	0.790	74.6	20.4	3	
80	04	14.81	112.4	95.9	+16.1	95.7	− 7.2	83.7	−12.0	166.4	118.1	0.776	77.7	21.5	4	Compare 95.
81	07 Oct.	15.89	112.8	90.8	+20.9	90.8	− 2.6	75.4	− 4.1	169.0	157.8	0.143	157.4	21.9	4	
82	09	15.90	113.3	46.4	− 1.4	43.6	−18.1	359.4	−17.6	139.1	151.3	0.230	144.9	22.4	4	
83	09	15.90	113.3	47.5	+13.1	48.7	− 4.5	8.7	− 4.8	19.5	165.6	0.062	173.5	22.4	5	Compare 73, 101.
84	09	15.90	113.3	49.5	+22.6	53.1	+ 4.2	14.9	+ 4.8	175.2	180.2	0.974	202.9	202.4	5	
85	09	15.90	113.3	59.8	+ 0.8	57.8	−19.3	18.4	−23.0	80.7	156.7	0.156	159.8	22.4	7	
86	09	15.90	113.3	60.4	+ 7.2	59.8	−13.2	23.8	−16.2	95.0	173.7	0.987	189.8	22.4	6	
87	09	15.90	113.3	76.3	+19.7	77.1	− 3.1	52.3	− 4.6	170.8	149.8	0.252	132.0	22.4	5	Compare 96, Z 156?
88	09	15.90	113.3	76.4	+ 1.1	75.4	−21.7	44.8	−31.0	122.4	142.4	0.372	127.1	22.4	4	
89	09	15.90	113.3	78.0	+16.0	78.4	− 7.0	54.1	−10.5	160.6	146.8	0.300	115.9	22.4	6	
90	09	15.90	113.3	80.4	− 2.7	79.3	−25.8	49.2	−37.8	120.1	134.8	0.501	112.0	22.4	4	
91	09	15.90	113.3	85.9	+32.2	86.0	+ 8.8	66.5	+13.9	160.4	134.2	0.513	110.7	202.4	4	
92	09	15.90	113.3	89.4	+14.0	89.4	− 9.4	72.0	−15.0	158.9	133.9	0.518	110.2	22.4	5	
93	09	15.90	113.3	90.6	+24.0	90.5	+ 0.6	74.6	+ 1.0	178.8	125.8	0.656	94.0	202.4	4	
94	09	15.90	113.3	93.6	+13.3	93.6	−10.1	79.2	−16.5	160.5	122.5	0.721	87.4	22.4	6	
95	09	15.90	113.3	97.9	+17.7	97.6	− 5.5	86.3	− 9.2	169.8	115.7	0.809	73.4	22.4	3	Compare 80.
96	04	16.80	114.4	77.4	+20.0	78.2	− 2.9	53.3	− 4.3	169.2	156.1	0.163	135.8	23.5	3	Compare 87.
97	04 Oct.	16.80	114.4	91.6	+17.7	91.5	− 5.7	75.4	− 7.6	170.3	127.7	0.628	98.9	23.5	7	Compare 115, 126.
98	03	18.83	115.7	84.2	+ 8.6	84.1	−14.8	60.4	−22.5	144.6	138.7	0.434	122.2	24.8	4	Compare 103+104, 134.
99	03	18.83	115.7	87.6	+13.3	87.7	−10.1	67.4	−15.8	157.3	135.0	0.498	114.8	24.8	5	Compare 111.
100	03	18.83	115.7	92.1	+13.6	92.1	− 9.8	75.0	−15.7	159.9	128.0	0.618	110.8	24.8	13	
101	01	18.81	116.1	47.8	+11.8	48.6	− 5.8	7.4	− 6.0	18.8	161.3	0.104	167.8	25.2	6	Compare 73, 83.
102	01	18.81	116.1	57.5	+ 0.2	55.3	−19.4	13.6	−21.5	62.9	155.7	0.169	156.5	25.2	5	
103	01	18.81	116.1	86.8	+10.8	86.8	−12.7	65.2	−19.6	151.0	136.2	0.476	117.7	25.2	5	} Combine.
104	01	18.83	116.2	84.6	+11.0	84.5	−12.4	61.4	−18.9	149.9	139.8	0.415	124.8	25.2	4	
105	01	18.83	116.2	91.2	+14.2	91.2	− 9.2	73.2	−14.7	160.5	130.4	0.578	106.0	25.2	5	
106	01	18.83	116.2	94.2	+11.4	94.2	−12.0	77.9	−19.4	156.0	124.8	0.671	94.9	25.2	7	
107	04	18.81	116.4	65.4	+25.9	68.0	+ 4.3	35.8	+ 5.7	150.8	168.3	0.047	182.1	205.5	6	
108	04	18.81	116.4	79.2	+28.6	80.4	+ 5.4	72.1	+ 8.6	168.2	132.8	0.537	111.0	205.5	4	A. P. 11; 2 and 4.
109	04	18.81	116.4	83.5	+18.2	83.8	− 5.3	61.1	− 8.0	166.4	143.6	0.351	132.6	25.5	4	[J.B.S.]
110	04	18.81	116.4	87.5	+14.4	87.6	− 9.1	66.9	−14.1	159.2	136.6	0.470	118.8	25.5	7	

TABLE OF RADIANTS.—*Continued.*

No.	G.M.T.1900+	L	a	δ	l'	b'	l	b	ι	η	q	π	Ω	s→	
111	04	18.82	116.4	88.9	+12.9	88.9	-10.5	69.0	-16.5	156.6	134.1	0.513	113.8	25.5	4	Compare 99. [J.B.S.]
112	04	18.81	116.4	90.0	+16.4	90.0	- 7.0	71.2	-11.2	164.6	133.2	0.529	111.9	25.5	6	[J.B.S.]
113	04 Oct.	18.82	116.4	90.8	+16.6	90.8	- 6.8	72.2	-11.0	165.1	132.3	0.544	110.2	25.5	5	
114	04	18.81	116.4	92.0	+15.5	92.0	- 7.9	74.5	-12.7	161.0	137.4	0.456	120.4	25.5	16	
115	04	18.81	116.4	93.4	+19.4	93.2	- 4.0	76.8	- 6.4	171.8	128.4	0.612	102.2	25.5	5	A.P. 11; 1 and 3. Compare 97, 126. [J.P.S.]
116	04	18.81	116.4	94.1	+25.0	93.7	+ 1.6	77.8	+ 2.6	176.7	127.6	0.624	100.8	205.5	5	
117	08	18.73	116.5	90.2	+14.3	90.2	- 9.2	71.2	-14.5	160.2	132.4	0.542	110.3	25.4	6	
118	03	19.82	116.6	91.5	+14.4	91.4	- 9.1	73.2	-14.5	158.7	136.5	0.472	118.7	25.7	23	
119	03	19.82	116.6	92.3	+11.0	92.3	-12.4	74.2	-19.7	154.4	128.6	0.608	102.9	25.4	4	
120	01	19.81	117.2	82.6	+11.7	82.6	-11.6	57.8	-17.2	149.4	134.5	0.335	135.2	26.2	5	
121	01	19.81	117.2	90.1	+14.4	90.1	- 9.0	70.7	-14.2	160.2	133.8	0.519	113.8	26.2	13	
122	01	19.81	117.2	91.9	+12.9	91.9	-10.6	73.5	-16.8	157.7	130.5	0.575	107.2	26.2	30+	
123	00	19.82	117.4	90.6	+15.9	90.6	- 7.6	71.5	-11.9	163.4	133.7	0.520	113.0	26.4	9+	
124	00	19.82	117.4	92.1	+15.0	92.4	- 8.5	74.5	-13.6	162.0	130.5	0.575	107.4	26.4	29+	
125	05	20.88	118.2	88.7	+16.2	88.7	- 7.2	67.9	-11.2	163.0	138.2	0.443	123.6	27.3	10	Compare 128+129, 130.
126	05	20.88	118.2	97.8	+19.2	97.4	- 4.1	82.7	- 6.6	172.0	124.4	0.678	96.0	27.3	5	Compare 97, 115.
127	09	22.92	120.3	76.3	+12.4	76.4	-10.4	45.9	-14.3	138.2	158.3	0.137	165.9	29.4	4	D, LXIV, 13*
128	09	22.92	120.3	89.4	+16.6	89.4	- 6.8	67.8	-10.6	163.3	140.4	0.404	130.2	29.4	7	} Combine. Compare 125, 130.
129	09 Oct.	22.92	120.3	90.5	+18.1	90.5	- 5.4	69.7	- 8.3	167.3	139.0	0.428	127.3	29.4	8	
130	05	23.86	121.1	90.3	+19.3	90.3	- 4.2	68.8	- 6.4	169.8	140.0	0.394	132.1	30.3	4	Compare 125, 128+129.
131	06	25.84	122.9	56.0	+12.6	56.6	- 7.0	24.6	- 7.3	24.6	162.2	0.926	186.5	32.0	4	
132	06	25.84	122.9	79.0	+28.5	80.3	+ 5.3	51.4	+ 7.4	158.4	159.3	0.124	170.6	212 0	5	
133	06	25.84	122.9	86.2	+17.6	86.3	- 5.8	61.0	- 8.6	162.7	149.9	0.251	151.7	32.0	6	
134	06	25.84	122.9	91.9	+ 7.7	92.0	-15.7	68.4	-24.0	143.1	137.3	0.457	126.6	32.0	6	Compare 98, 103 + 104,
135	06	25.84	122.9	100.0	+33.7	98.4	+10.5	80.6	+16.8	158.0	129.3	0.596	110.5	32.0	5	
136	06	25.84	122.9	102.0	+24.1	101.0	+ 1.1	85.6	+ 1.8	177.8	143.6	0.643	139.2	212.0	6	
137	06	26.86	123.9	102.4	+15.3	102.1	- 7.6	86.5	-12.3	164.8	54.3	0.657	321.6	33.0	5	
138	06	26.86	123.9	113.8	- 6.8	117.0	-28.0	68.8	-42.9	122.2	126.4	0.643	105.9	33.0	4	
139	01 Nov.	13.84	142.0	102.3	+35.0	100.3	+12.0	60.6	+16.8	118.3	160.8	0.107	192.7	231.3	4	
140	01	13.84	142.0	109.0	+67.6	100.0	+44.8	86.5	+50.8	115.2	121.1	0.725	113.5	231.3	4	
141	01	13.84	142.0	110.6	+ 2.8	111.8	-19.1	81.7	-29.3	132.1	138.8	0.430	148.8	51.7	6	
142	01	13.84	142.0	159.2	+52.6	137.8	+40.0	130.3	+67.0	112.6	94.3	0.894	59.8	231.3	4	Compare 148, 157.
143	01	13.84	142.0	169.4	+22.7	161.2	+16.6	355.4	+27.2	31.7	119.5	0.749	110.3	231.3	4	
144	03	148.4	142.4	76 2	+31.0	78.1	+ 8.1	37.7	+ 8.7	32.1	163.5	0.080	198.7	231.8	5	
145	03	14.84	142.4	94.4	+ 8.0	94.5	-15.4	60.0	-20.9	110.9	22.5	0.144	276.5	51.6	4	
146	09	14.84	143.1	128.6	- 2.0	131.6	-20.0	121.8	-33.6	140.0	120.6	0.733	113.6	52.3	2	Compare 163.
147	09	14.84	143.1	139.9	+34.8	131.4	+18.3	121.8	+30.7	147.6	107.6	0.899	87.5	232.3	4	Compare 176, Z167?
148	09	14.84	143.1	155 0	+58.0	138.5	+43.5	127.6	+72.4	107.0	94.4	0.993	61.2	232.3	4	Compare 142, 157.
149	04	14.88	143.3	137.8	+18.2	134.8	+ 1.9	128.7	+ 3.2	176.7	103.9	0.932	80.3	232.6	5	
150	04	14.88	143.3	138.0	+49.2	124 7	+31.4	103.9	+50.7	122.6	113.3	0.834	99.4	232.5	5	Compare 158, 160.
151	04	14.88	143.3	147.2	+60.8	124.8	+44.1	86.8	+68.6	102.4	162.4	0.898	207.4	232.6	5	
152	03	15.83	143.5	72.5	+40.5	76.1	+17.9	32.4	+18.1	43.2	153.0	0.203	178.9	232.8	3	Compare 155.
153	03	15.83	143.5	87.0	+22.8	87.2	- 0.6	51.2	- 0.7	24.9	178.2	0.001	229.2	52 8	4	D, LXXV*
154	03	15.8	143.5	112.0	+10.8	112.0	-11.0	89.2	-16.8	153.0	140.4	0.402	53.6	52,8	4	
155	01	15.87	144.0	70.2	+41.5	74.4	+19.1	29.7	+18.6	40.2	150.3	0.242	173.9	233.2	4	Compare 152.
156	01	15.87	144.0	137.3	+19.2	134.0	+ 2.4	127.0	+ 4.1	174.9	126.8	0.633	127.0	233.3	2+	
157	01	15.87	144.0	155.4	+50.9	136.2	+37.4	124.2	+62.4	116.3	98.8	0.966	70.8	233.3	5	Compare 142, 148.
158	00	15.84	144.2	138.8	+47.8	125.4	+48.9	105.8	+48.9	124.6	113.7	0.829	100.9	233.5	2	Compare 150, 160.
159	09	16.00	144.2	147.0	+36.5	136.6	+21.8	129.8	+36.9	142.3	100.9	0.953	75.3	233.5	5	
160	06	16.82	144.8	137.0	+49.0	124.0	+31.0	111.2	+49.4	122.2	116.3	0.797	106.6	234.1	3	Compare 150, 158.
161	06 Nov.	16.83	144.8	78.6	+29.0	80.0	+ 6.0	249.9	+ 6.4	22.2	17.0	0.850	268.2	234.1	7	D, LXVII, 26*
162	06	16.82	144.8	101.0	+34.9	99.2	+11.8	67.1	+15.9	128.3	159.6	0.121	193.2	234.1	5	
163	06	16.83	144.8	130.7	- 1.2	133.5	-18.7	124.1	-31.4	143.0	120.0	0.741	114.2	54.1	4	Compare 146.
164	06	16.82	144.8	140.0	+26.7	130.1	+10.6	119.2	+17.7	142.8	149.9	0.249	173.8	234.1	4	
165	04	18.67	146.5	95.4	+ 10.9	95.4	-12.4	60.2	-15.8	105.8	16.5	0.079	268.4	235.6	4	

TABLE OF RADIANTS.—*Continued.*

No.	G.M.T.1900+	L	α	δ	l'	b'	l	b	ι	η	q	π	Ω	s→	
166	10 May 4.97	313.3	334.0	− 3.4	334.6	+ 6.9	349.8	+11.2	166.2	55.0	0.677	155.1	44.1	6+	η Aquarids.
167	10 May 6.93	315.4	337.7	− 0.6	339.2	+ 8.2	356.0	+13.0	163.2	51.2	0.607	148.3	46.0	25	η Aquarids. [G.H.]
168	10 May 11.99	320.1	342.0	− 0.6	343.2	+ 6.5	359.5	+10.5	166.7	52.1	0.630	155.1	50.8	5+	η Aquarids.
169	10 Aug. 1.93	39.7	14.6	+16.3	19.8	+ 9.3	5.9	+14.9	162.3	122.0	0.935	13.3	129.2	3	
170	10 1.93	39.7	16 9	+ 8.0	18.6	+ 0.8	4.5	+ 1.2	178.5	124.7	0.686	18.5	129.2	6	
171	10 6.9	44.5	10.1	+19.4	17.0	+13.8	356.6	+21.6	149.7	133.2	0.538	30.4	134.0	3	
172	10 6.9	44.5	30.4	+56.4	51.0	+44.7	62.8	+67.4	111.4	82.9	0.998	299.8	134.0	5	D, XXXII, 5. Compare 175.
173	10 11.86	49.3	7.2	+58.9	38.0	+49.4	9.3	+75.3	101.4	99.3	0.987	337.3	138.8	2	Compare 51, 57.
174	10 11.86	49.3	7.6	+68.6	48.9	+56.6	230.4	+75.8	75.8	90.4	1.013	319.6	138.8	3	
175	10 11.86	49.3	35.8	+56.6	54.6	+39.6	63.6	+65.9	113.4	84.0	1.003	306.7	138.8	4	Compare 53, 175.
176	07 Nov. 14.88	142.6	139.5	+35.0	131.0	+18.3	121.5	+30.8	147.6	107.4	0.900	86.7	231.9	3	Compare 147, Z 173 ? ?

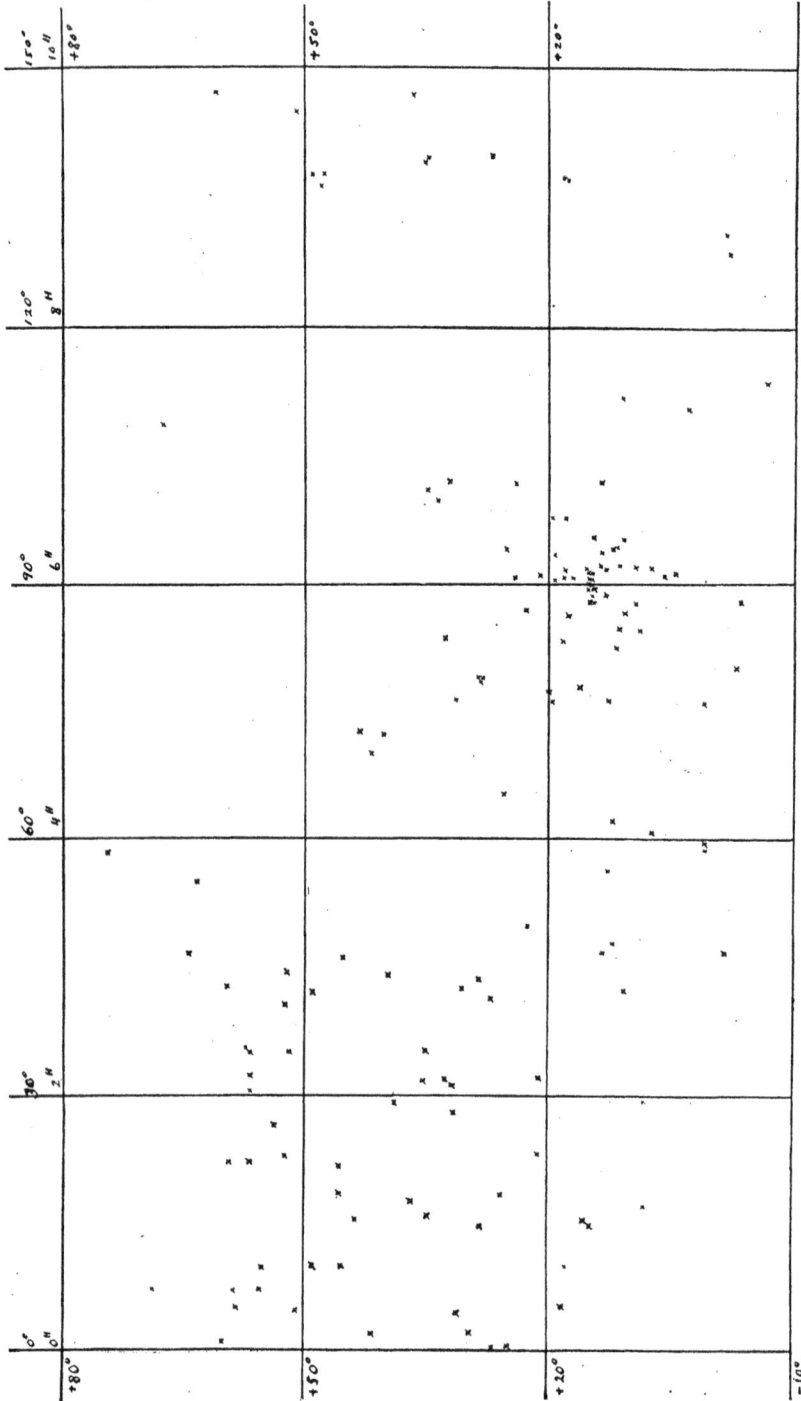

Fig. 2.

UNCERTAIN RADIANTS.

1900+			L	α	δ	$\xrightarrow{s}$	Notes.
05	Jan.	1.80	191.3	175.9	+37.5	4	D CXXXVIII ?
09	"	18.83	208.9	198.2	− 4.0	4	D CLVIII ?
04	Apr.	18.75	297.4	279.1	+30.8	3	
04	"	18.75	297.4	293.6	+58.0		D CCXXIV ?
01	Aug.	8.79	46.5	19.4	+15.6	3	2 on Aug. 8.
01	"	9.77	47.5	10.2	+26.6	5	
08	Oct.	18.73	116.5	43.3	+ 3.8	3	D XXXVIII, 6, 7
08	"	18.73	116.5	106.5	+33.8	2	Z 161?
04	"	18.81	116.4	100.8	+30.3	2	D LXXXIX.
03	"	18.83	115.7	61.7	+24.4	2	D LXIII, 13*
03	"	19.82	116.6	48.2	− 4.7	3	D XLI, 5.
03	"	19.82	116.6	61.5	+23.3	2	D LXIII, 13*
06	"	25.84	122.9	64.1	−16.3	3	
00	Nov.	13.95	142.3	93.8	+44.0	3	
01	"	13.84	142.0	97.8	+16.5	3	
01	"	13.84	142.0	135.4	+ 6.2	3	
03	"	14.84	142.4	118.4	+55.8	3	
03	"	14.84	142.4	147.7	+53.6	2	
01	"	15.87	144.0	135.7	− 4.7		
01	"	15.87	144.0	145.2	+70.6	2	
00	"	15.84	144.2	92.2	+24.0	4	Nov. 13, 1 seen, Nov. 15, 1 seen.
00	"	15.84	144.2	93.8	+44.0	3	
00	"	15.84	144.2	158.4	+54.3	3	Nov. 13, 1 seen.
09	"	16.00	144.2	120.9	+11.8	3	
09	"	16.00	144.2	130.9	+ 8.5	3	
06	"	16.83	144.8	114.2	− 9.0	3	
1899 Nov. 24.				20.2	+37.0	8	Beilids. 75 meteors seen in all between 8ʰ10ᵐ and 10ʰ10ᵐ.
				23.6	+29.9	6	
				25.2	+40.8	5	

TABLE OF INCLINATIONS AND PERIHELIA.

Inclinations of Orbits.	0° to 10°	10° to 20°	20° to 30°	30° to 40°	40° to 50°	50° to 60°	60° to 70°	70° to 80°	80° to 90°	90° to 100°	100° to 110°	110° to 120°	120° to 130°	130° to 140°	140° to 150°	150° to 160°	160° to 170°	170° to 180°	T'l
Before combination	0	3	6	5	3	2	3	3	2	8	14	17	16	12	20	16	33	12	175
After combination	0	1	5	5	1	2	3	3	2	7	11	15	14	11	14	14	22	9	139
Zezioli's	4	22	9	17	15	18	22	12	3	23	7	13	8	5	3	3	2	0	189

Longitudes of Perihelia.	0° to 10°	10–20	20–30	30–40	40–50	50–60	60–70	70–80	80–90	90–100	100–110	110–120	120–130	130–140	140–150	150–160	160–170	170–180	180–190	190–200	200–210	210–220	220–230	230–240	240–250	250–260	260–270	270–280	280–290	290–300	300–310	310–320	320–330	330–340	340–350	350° to 360°	T'l
Before combination	2	3	1	2	2	2	4	6	6	6	13	20	4	6	4	5	3	5	4	3	2	1	2	0	2	0	2	1	4	6	4	7	8	13	7	5	175
After combination	2	3	1	2	2	1	4	4	4	4	6	12	4	6	4	5	3	5	4	3	2	1	2	0	2	0	2	1	4	6	4	7	8	10	6	5	139
Zezioli's	2	3	2	3	2	2	1	1	5	2	4	2	2	3	4	4	5	4	6	3	8	7	8	3	6	10	5	10	11	5	12	10	10	8	9	7	189

MAGNITUDES OF METEORS.

Year.	>0	0	1	2	3	4	5	6	<6	—	Total
1898		1								119	120
1899	0	1	10	12	4	7	1	1	0	101	137
1900	21	2	41	54	59	39	31	7	0	182	436
1901	10	14	46	96	128	116	51	15	0	50	526
1902	2	11	19	44	91	51	12	2	0	12	244
1903	28	33	71	122	268	166	34	2	0	7	731
1904	12	24	101	125	273	170	48	7	0	27	787
1905	8	10	27	45	123	106	33	4	0	13	369
1906	15	6	27	52	111	107	30	22	2	3	375
1907	8	8	15	27	64	95	31	8	0	2	258
1908	5	2	10	20	27	34	14	1	0	13	126
1909	15	16	37	117	363	410	197	27	5	30	1213
1910	10	5	16	71	181	266	104	16	0	7	676
1900 } 1908 }	109	110	357	585	1,142	884	284	68	2	309	3,852
1909 } 1910 }	25	21	53	188	544	676	301	43	5	37	1,889
Others } 1898–9 }	0 0	0 2	3 10	14 12	14 4	12 7	12 1	4 1	10 0	47 } 220 }	373
Total											6,114

MAGNITUDES OF METEORS. PERCENTAGE.

Year.	>0	0	1	2	3	4	5	6	<6
1900 } 1908 }	03.1	03.0	10.3	16.5	32.3	25.0	08.0	01.9	00.1
1909 } 1910 }	01.3	01.1	02.9	10.2	29.4	36.5	16.3	02.3	00.3

COLORS OF METEORS.

Year.	Red.	Orange.	Yellow.	Green.	Blue.	Purple.	White.	Total.
1898				1				1
1899	14		1	18	1		1	35
1900	81		13	53	5		10	162
1901	96	67	15	77	2		27	284
1902	38	40	1	37	4			120
1903	108	178	17	89	6	4	67	469
1904	90	27	58	24	3	7	30	239
1905	32	2	28	11	2	1	1	77
1906	42	15	17	17		2	17	110
1907	46	5	12	7	3		8	81
1908	20	6	2	2	1		6	37
1909	120	3	96	36	4	1	16	276
1910	51	3	50	11	1	3		119
Total	738	346	310	383	32	18	183	2,010
Per cent.	36.7	17.2	15.4	19.1	01.6	00.9	09.1	100.0

DURATIONS OF FLIGHT WITH REGARD TO COLOR.

Seconds	0.1	0.2	0.3	0.4	0.5	0.6	0.7	0.8	0.9	1.0	1.1	1.2	1.3	1.4	1.5	1.6	1.7	1.8	2.0	2.5	3.0	3.5	4.0	5.0	
Red.....	4	11	60	101	75	34	12	12	4	11		5	1		5			·1	3	4	·1			1	1902–1908
Orange ..	2		57	103	40	25	5	7		8		2			3	1		3	2	1	1		1		
Yellow...		2	19	58	18	12	·1	1		2		1					1			1					
Green ...		2	24	54	38	28	6	2	1	8		1			3		1	1	5			1	1		
Blue.....			2	1	3	1		2		1					2	1			1	1					
Purple...	1	1	4	1	3				1		1	1					1								
White ...		1	8	33	17	10	1	3	1	2		3													

Seconds	0.1	0.2	0.3	0.4	0.5	0.6	0.7	0.8	0.9	1.0	1.1	1.2	1.3	1.4	1.5	1.6	1.7	1.8	2.0	2.5	3.0	3.5	4.0	5.0	
Red.....	3	3	43	60	22	12	3	3		2			2		1			1	1	1	1				1909+1910
Orange ..				3	1	1			1	1															
Yellow...		1	32	56	30	7		2				1						1	2						
Green ...		3	5	15	5	3		2		4		2	1		2			1	1						
Blue.....				1	1											1		1						1	
Purple...			2									1						1							
White ...			4	5	1								1												

Means	All 1902–'08	No.	All 1909+'10	No.	<1s.1 1902+'08	No.	<1s.1 1909+'10	No.	All 1902–'10	No.	<1s.1 1902–'10	No.	Δ
	s		s		s		s		s		s		s
Red.....	0.557	345	0.479	158	0.468	324	0.414	151	0.532	503	0.451	475	0.091
Orange ..	0.529	261	0.600	7	0.448	247	0.600	7	0.531	268	0.452	254	0.079
Yellow...	0.474	116	0.464	132	0.434	113	0.414	128	0.469	248	0.423	241	0.046
Green ...	0.605	176	0.661	44	0.489	163	0.486	37	0.616	220	0.488	200	0.128
Blue.....	1.053	15	1.920	5	0.570	10	0.550	2	1.245	20	0.567	12	0.678
Purlpe...	0.671	14	1.000	4	0.482	11	0.400	2	0.745	18	0.469	13	0.276
White ...	0.506	79	0.569	13	0.478	76	0.517	12	0.514	92	0.484	88	0.030
———	0.404	1319			0.398	1313							

EXCEPTIONAL METEORS.

Year.	Sta. Meteors.	Var. Magn.	Irr. Path.	Curved Path.	Rem. Trains.	Hazy.	Trains 2s or over.
1898	0	0	0	—	0	0	—
1899	0	—	—	—	—	—	—
1900	0	1	3	4	—	2	7
1901	1	2	3	5	2	1	20
1902	1	—	—	1	—	1	10
1903	1	1	1	2	2	4	50
1904	1	4	5	1	—	—	51
1905	0	—	2	—	—	—	21
1906	0	—	—	1	2	—	24
1907	0	1	—	—	1	2	12
1908	1	1	4	—	—	1	4
1909	0	3	—	5	3	4	66
1910	0	2	1	—	1	2	32
Total	5	15	19	19	11	18	297

EXPLANATION OF TABLES.

Table of Radiants.—This table contains the parabolic elements calculated for the 175 radiants, which were considered good enough to justify the computations, and which did not belong to the Perseids or Leonids. The columns give from left to right: (1) Serial number of radiant, (2) Greenwich Mean Time of observations, (3) longitude of the meteoric apex, (4) right ascension of radiant, (5) declination, (6) apparent longitude, (7) apparent latitude, (8) true longitude, (9) true latitude, (10) inclination to the ecliptic of parabolic orbit, (11) the angle between the true position of the radiant point and the sun, (12) perihelion distance in terms of earth's distance from sun, (13) longitude of perihelion, (14) longitude of ascending node, (15) number of meteors from which radiant was deduced, (16) notes and references.

In (16), when the radiant was observed by anyone other than myself, the observer's initials are enclosed in square brackets. Other references are as follows: A. P. — — refers to *Annales de l'Observatoire d'Athénes*, Vol. III, with page and number of the radiant on page.

D, — — refers to W. F. Denning's "General Catalogue of Meteor Radiants," giving the group and number in group. An asterisk following the number means that several nights' observations were used by observers referred to.

Z, — — refers to Schiaparelli's corrected orbits, deduced from Zezioli's observations, giving the number of the orbit in his work.

Uncertain Radiants. — This table gives 26 uncertain radiants. The columns give from left to right: (1) Greenwich Mean Time, (2) longitude of meteoric apex, (3) right ascension of radiant, (4) declination, (5) number of meteors from which radiant was deduced, (6) notes and references.

Table of Inclinations and Perihelia.—The first of these tables gives for each 10° of inclinations three series of results. The first row contains the inclinations of all 175 orbits, just as taken from Table of Radiants. The second row gives the distribution after allowance has been made for the same meteor stream reappearing one or more times. The third row gives the distribution as taken from Schiaparelli's orbits, based on Zezioli's observations.

The second table gives the longitudes of perihelia, in three exactly similar rows.

Table of Magnitudes of Meteors.—This table gives the number of meteors of each magnitude observed in every year. Below are three combinations, first all meteors seen from 1900 to 1908 inclusive, second all meteors seen in 1909 and 1910, third for al other years.

Percentage.—This table gives the percentage of each magnitude in the first and second combination just mentioned. Of course no account is taken of meteors whose magnitudes are not given, in this table.

Table of Colors of Meteors.—This table is self-explanatory.

Table of Duration of Flight, etc.—Two tables are given under this head. The first is divided into two periods, namely 1902 to 1908 inclusive, 1909 and 1910. In each portion the numbers of meteors of each color are divided up to show how many of a given duration of visibility were seen.

The second table gives the mean visibility for each color divided as follows: (1) All meteors, 1902–08, (2) number of meteors, (3) all meteors 1909+1910, (4) number of meteors, (5) meteors whose length of visibility was not over 1ˢ.0 in 1902–1908, (6) number of these meteors, (7) same for 1909+1910, (8) number of these meteors, (9) all meteors 1902 to 1910 inclusive, (10) their number, (11) meteors not over 1ˢ.0, (12) their number, (13) difference between (9) and (11).

Table of Exceptional Meteors.—This gives the peculiar meteors seen in each year under the following headings: (1) Stationary meteors, (2) meteors whose magnitude varied, (3) meteors with irregular paths, (4) meteors with curved paths, (5) very remarkable trains left, (6) meteors which certainly had hazy nuclei,(7) numbers of trains recorded as being visible at least 2.0 seconds.

This table is only partially correct because undoubtedly not all peculiar meteors seen were so recorded.

DEDUCTIONS FROM TABLES.

When my inclinations and perihelia are examined, they at once show certain marked maxima. As to the inclinations very many more are retrograde than direct with a strong maximum at 160°–170°. This is worthy of attention when we remember that most of the short period comets move with direct motion. However, Zezioli's orbits show more of direct than retrograde motion. I can, at present, only explain this difference in our results by pointing out that while his observations were made throughout the entire year, most of these were made between July 20 and November 20, and therefore are not well enough distributed to be strictly comparable with his. These were also generally made after midnight. As to the perihelia two maxima are shown, one about 110° and the other at 320°. This last agrees fairly well with Zezioli, but no maximum near the first position is shown by his orbits.

In the results from the tables of magnitudes the most important is that which shows that during 1900 to 1908 inclusive, when observations were made mostly in Virginia, at the Leander McCormick Observatory, meteors of the third magnitude were the most numerous by far. However, during 1909 and 1910, when observations were made in California, at the Lick Observatory, meteors of the fourth magnitude were greatly in the majority. This proves conclusively, since the numbers are quite comparable, that there is every reason to believe we would find meteors of the fifth magnitude, and so on indefinitely, most numerous could we get sky as much clearer than the Lick, as the Lick is clearer than the Virginia sky.

The series representing meteor magnitudes is similar to that for stars, only the factor seems less than 2.5.

The question of the length of visibility of meteors, with regard to their color appeared a sufficiently interesting question to investigate. Consequently the two tables on p. 31 were formed for this purpose.

It was not until 1902 that this particular datum was observed, which explains why meteors seen previously are not included.

To understand the results some remarks are necessary, In the first place few meteors whose magnitude was below the third had color recorded for them. Yellow and orange were used loosely, and could often have been interchanged. Blue, purple and to a lesser extent green meteors are difficult to distinguish from white meteors, unless bright or slow. Indeed very few blue or purple meteors were seen. Finally, white practically means no color could be detected, and as a rule was not entered on the observing book, which will explain why there are so few white meteors in the lists.

The first part of the table merely gives the data for forming the second. In this latter the meteors are studied in three classes, each with two subdivisions. All those observed 1902 to 1908 inclusive form one class, those observed 1909 and 1910 at the Lick Observatory form the second, and these are combined to form the third class and give the definitive results.

Each class is subdivided to give the means for all meteors and then the means for those whose visibility did not exceed 1.0 second. This last is the column which should be studied to obtain results, because any meteor whose visibility is over 1.0 second is an unusual one. The results are rather surprising, though well marked.

Yellow meteors have the shortest, orange and red the next and equal times of visibility, finally green and white almost equal and the longest.

Blue and purple are also visible a longer time, but the results depend on too few meteors to have much weight. It is noteworthy that the other columns, in general, bear out the results deduced from the last. Taking as a first assumption that meteors of all colors enter the atmosphere with the same mean velocity and become visible at the same mean height, then we must conclude that yellow meteors are composed of materials which are more inflammable than red and orange meteors, and these in turn more so than white or green. It may be objected that the error of observation is so great that the differences in the column referred to mean little or nothing. While it is true enough that for any single meteor the error is large, yet when the means of several hundred are taken, as in this case, I believe the accidental errors are largely eliminated.

Finally the mean values for meteors whose color was unrecorded in 1902–1908 are given, merely for comparison. As most of these were of the fainter magnitudes than the third, it is quite obvious that they should in general be seen a less time than brighter meteors, as indeed the table proves.

The exceptional meteors call for no further comment here than to say that my observations give fewer per thousand than is usually the case. As there were 297 trains of 2.0 seconds duration, or over, it seems that about one meteor in 20 leaves such an one. A full description of all these exceptional phenomena will probably be published later in another paper.

CONDENSED SUMMARY OF RESULTS.

This summary is intended to give in a very few words what appear to be the main results deduced in this paper.

1. Stationary radiants appear to be rare if they exist at all.

2. Proof that Halley's Comet and the η Aquarids are intimately connected.

3. The change in position of the Perseid radiant, from day to day, is fully confirmed.

4. The Orionids do not seem to have a stationary radiant.

5. The radiant of the Leonids shows no apprecialbe change of position from day to day.

6. The existence of the so-called $\alpha - \beta$ Perseids, except in August, is not confirmed.

7. By observing in a clearer atmosphere, meteors of the fourth magnitude are in the majority, while formerly more of the third magnitude were seen.

8. Yellow meteors have the shortest time of visibility, red and orange somewhat longer times, while green and white are seen longest.

9. Peculiar meteors are not so common as thought.

LEANDER McCORMICK OBSERVATORY,
    UNIVERSITY OF VIRGINIA.